雁荡山联合国教科文组织世界地质公园

科学考察指南

FIELD TRIP GUIDE OF YANDANGSHAN UNESCO GLOBAL GEOPARK

袁茂珂　杨　林　卢琴飞　等 编著

中国地质大学出版社
CHINA UNIVERSITY OF GEOSCIENCES PRESS

图书在版编目(CIP)数据

雁荡山联合国教科文组织世界地质公园科学考察指南/袁茂珂等编著. —武汉:中国地质大学出版社,2021.8

ISBN 978-7-5625-4972-7

Ⅰ.①雁…

Ⅱ.①袁…

Ⅲ.①雁荡山-地质调查-指南

Ⅳ.①P562.55-62

中国版本图书馆CIP数据核字(2021)第137373号

雁荡山联合国教科文组织世界地质公园

科学考察指南 袁茂珂 杨 林 卢琴飞 等 编著

责任编辑:胡珞兰 龙昭月 选题策划:胡珞兰 毕克成 段 勇 责任校对:徐蕾蕾

出版发行:中国地质大学出版社(武汉市洪山区鲁磨路388号) 邮政编码:430074

电 话:(027)67883511 传 真:(027)67883580 E-mail:cbb@cug.edu.cn

经 销:全国新华书店 http://cugp.cug.edu.cn

开本:880毫米×1230毫米 1/24 字数:123千字 印张:4.25

版次:2021年8月第1版 印次:2021年8月第1次印刷

印刷:湖北新华印务有限公司 印数:1—5300册

ISBN 978-7-5625-4972-7 定价:38.00元

《雁荡山联合国教科文组织世界地质公园科学考察指南》编委会

FIELD TRIP GUIDE OF YANDANGSHAN UNESCO GLOBAL GEOPARK EDITORIAL COMMITTEE

目 录
Contents

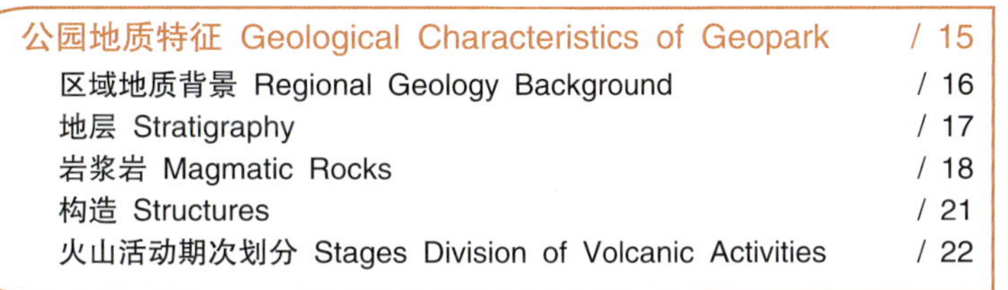

公园基本情况

General Information of Geopark

公园简介

雁荡山联合国教科文组织世界地质公园(以下简称雁荡山世界地质公园)位于浙江省温州市和台州市内,由雁荡山、楠溪江、方山-长屿硐天3个各具特色的园区组成,面积298.80平方千米,是以火山岩地质地貌为主导,千年宗教历史、山水文化和石文化交相辉映的综合性自然公园。

雁荡山属大型滨海山岳型风景名胜区,是亚洲大陆边缘巨型火山(岩)带中白垩纪火山的典型代表,是研究流纹岩的天然博物馆。雁荡山的一山一石记录了亿年前的一座复活型破火山演化的历史。

公园的锐峰、叠嶂、方山、怪洞、门阙造型奇特,神韵优美,意境深邃,无不令人惊叹。方山顶上有秀丽的云海佛光、日月同升和瀑布倒流景色。沈括、谢灵运、徐霞客、莫言等数百位古今名人在此留下了足迹,赋予雁荡山"寰中绝胜""天下奇秀"之赞誉。雁荡山不附五岳、不类他山而有独特的品格,"日景耐看,夜景销魂""一景多变,变幻造景""观山景,品海鲜"。古人云:不游雁荡是虚生;今人云:不游夜雁荡是虚生。

楠溪江在公园内柔曲摆荡,急缓有度,两岸奇岩峭壁、仙桥怪洞、飞瀑深潭。中游村落星罗棋布,苍坡、岩头、芙蓉等古村落布局讲究、古色古香,孕育了光辉灿烂的永嘉文化。秀丽的自然风光、浓厚的耕读文化和永嘉昆曲等共同组成了公园千年文化脉搏。

Introduction of Geopark

Yandangshan UNESCO Global Geopark (hereinafter refferred to as Yandangshan Global Geopark) is located in the cities of Wenzhou and Taizhou in Zhejiang Province. It comprises Yandangshan, Nanxijiang, Fangshan-Changyu Dongtian (Changyu Cave) with a total area of 298.80km^2. It is dominated by volcanic geology and landscape, integrated with religious history, cultural landscape and stone culture to form a comprehensive nature park.

Yandangshan, a large-scale coastal and mountains scenic area, is a representative of Cretaceous volcanoes found in the huge volcanic belt on the continental margin of Asia. It is a natural museum of rhyolites. Each mountain and each stone of Yandangshan recorded the evolution of a revived caldera hundreds of millions years ago.

The peaks, cliffs, lava plateaus, caves, stone gates in the geopark with peculiar shapes, elegant charm and abstruse conception form a breathtaking landscape. The beautiful scenery of Sea of Clouds, Buddha's light, Simultaneous Rising of Sun and Moon as well as Reversing Falls can be viewed on the top of Fangshan. Hundreds of ancient and modern celebrities such as Shen Kuo, Xie Lingyun, Xu Xiake, and Mo Yan had left their footprints here and had entitled Yandangshan as "the unique spot under heaven" and "wonders of the world". Being different from the "Five Mountains"and other mountains, Yandangshan possesses unique style. Its tour planning includes"presentable and fascinating day and night view" "diverse landscapes from different views" as well as "mountain scenery and seafood appreciation". Our forefathers have said it would be regretful for not visiting Yandangshan; while the moderns have said it would be regretful for not visiting the night view of Yandangshan.

The Nanxijiang flows through the geopark with peculiar rocks, cliffs, mysterious bridges, grotesque caves, plunging waterfalls and deep pools on both banks. The villages in mid-stream scatter all over like stars in the sky. Those ancient villages like Cangpo, Yantou and Furong specifically are famous for the layout with antique beauty which has nourished fabulous Yongjia Culture. Thousand years of geopark culture highlights splendid natural scenery, vibrant farming-studying culture and Yongjia Kun Opera, etc.

公园峰丛
Peak Cluster of Geopark

观夕硐　Guanxi Cave

During Cretaceous Period, volcanic eruption produced ash fall which was then settled to form thick layers of tuff. Since it was easy to excavate and process, quarry started operating in Northern and Southern Dynasties. In past over 1500 years, stonemasons have excavated 28 cave groups with 1314 smaller caves on mountains. The caves vary in shapes and forms. Some are isolated, connected, ring-shaped or piled up, while others are standing side by side filled with twists and turns. Therefore, the caves have been complimented as "naturally created man-made caves". Some caves have water seeping out from the walls. The colored water creates natural pictures on the wall.

Yandangshan, epitomizing landscape aesthetics, natural science and history & culture, is one of the most famous "three hills and five mountains" mountains in China. It can serve for sightseeing, going on holidays, field trips, scientific popularization education, cultural tracing-back, and religious pilgrimage. It is also world-class precious inheritance heritage.

公园在白垩纪火山爆发，空落火山灰构成的巨厚凝灰岩，由于岩石性质适宜开凿，便于加工，成为自南北朝以来历史悠久的古代采矿遗址。1500多年以来，石匠们依势取石留下了28个硐群，1314个硐窟。硐群形态各异，组合形式多样，有的孤立，有的串联，有的环生相叠，有的几硐并峙，曲折回绕，幽深莫测，被赞叹为“虽由人作，宛自天开”。有的硐壁裂缝渗水织彩，形成各式各样的天然壁画。

雁荡山集山水美学、自然科学、历史文化于一身，居身“三山五岳”之中，兼备观光旅游、休闲度假、科学考察、科普教育、文化追踪、宗教朝虔于一体，为世界级的宝贵遗产。

楠溪江　Nanxijiang

公园位置

雁荡山世界地质公园东临乐清湾，距温州市70千米，距杭州市300千米。地理坐标：东经120°41′40″—121°27′40″，北纬28°12′30″—28°30′00″。

公园位于我国东南沿海经济发达地区，北接长三角经济圈，南与福州、厦门、汕头经济带连成一线。区位交通极为便捷，已纳入上海、南京等城市的3小时交通圈，甬台温高速公路、高铁、104国道直达公园。途经本地的杭甬温高速、高铁、铁路和温州机场、港口构成通达全国乃至世界的海陆空快捷交通网络。

Location of Geopark

Yandangshan UNESCO Global Geopark, facing Yueqing Bay in the east, is 70km away from Wenzhou City and 300km away from Hangzhou City. The geographical coordinates are: E 120°41'40"-121°27'40" and N 28°12'30"-28°30'00".

The geopark is located at economically developed coastal areas of Southeast China, bordering Yangtze River Delta Economic Circle in the north and lining up with Fuzhou, Xiamen and Shantou in the south. In this region, the transportation is extremely convenient. It has a 3-hour transportation circle including cities like Shanghai and Nanjing as well as direct transport links such as the Ningbo-Taizhou-Wenzhou Expressway, high-speed rail lines and National Road 104. Besides, the Hangzhou-Ningbo-Wenzhou Expressway, high-speed rail lines, railways as well as Wenzhou Airport and port via local areas have constituted a convenient sea-land-air transportation network connected to all parts of the country and the world.

雁荡山世界地质公园在浙江省的位置示意图
Location of Yandangshan UNESCO Global Geopark in the Map of Zhejiang Province

地貌特征

雁荡山系括苍山南缘的一支余脉，地貌属中低山区。山势呈北东-南西走向，最高峰海拔1108米，向东、西两侧海拔降低。中山主要分布在雁湖、百岗尖、乌岩尖一带，岩性为石英正长岩。中低山处山体与沟谷相间分布，河谷 - 盆地串珠相连，在楠溪江流域段形成河谷冲积平原、洪积平原。

Geomorphology

Yandangshan is one of the southern ranges of Kuocang Mountain with a middle to low mountainous terrain. It runs in northeast-southwest direction with the highest point at 1108m. The mountain is lower in the east and west sides. The central part of the mountain is located at Yanhu Lake, Baigangjian and Wuyanjian areas with quartz syenite as the major rock type. The middle-low part of the mountain alternates with the valleys and closely linked with basins to form alluvial and diluvial plains along the Nanxijiang.

公园河流地貌
River Landform in Geopark

公园火山岩地貌 Volcanic Landform in Geopark

气候特征

公园为亚热带海洋性季风气候，具有热量丰富、光能充足、温湿多雨、四季分明、水热同步的气候特点。年均平均温度13.5℃，7月份平均温度27℃，1月份平均温度 5~7℃，全年平均湿度为77%，平均无霜期269天。

公园年平均降雨量1 935.6毫米，区内暴雨中心地多年平均降雨量高达2127毫米。降雨量季节分配明显，主要集中在梅雨期和夏秋季，其中受台风影响尤其显著。公园内的水系属山区水系，河流多为源短流量小的溪坑，受大气降水的影响较大。

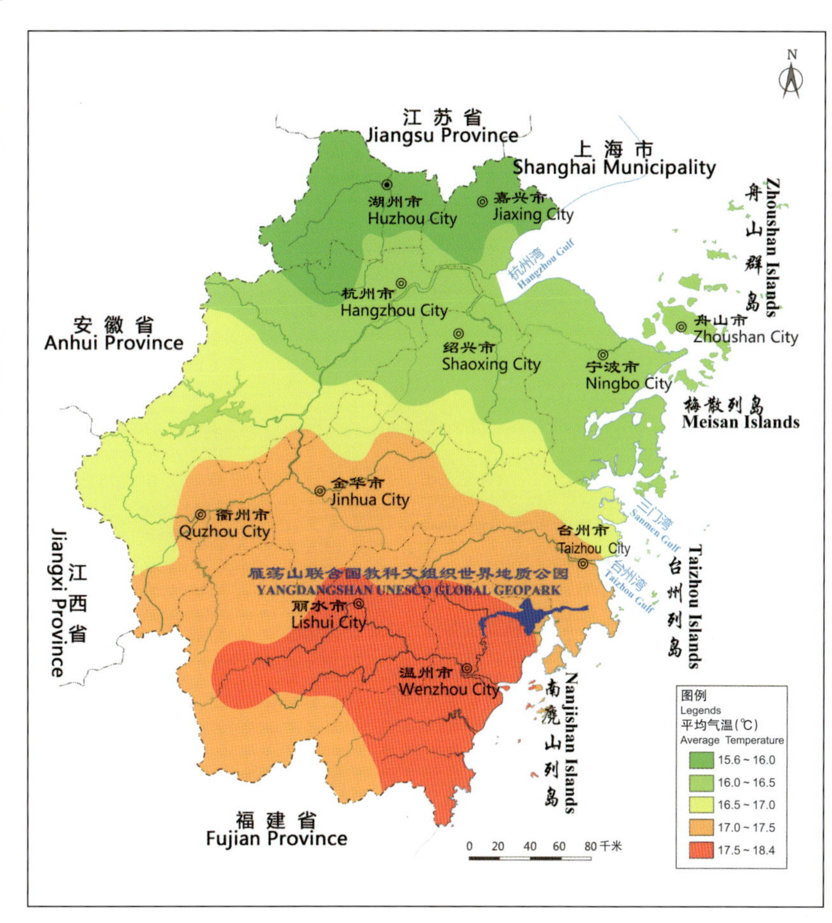

浙江省年平均气温图
Average Annual Temperature of Zhejiang Province

Climate

The climate of the geopark belongs to subtropical warm maritime monsoon climate with plenty of sunshine and four distinctive seasons. It is humid and rainy where the rainy season is synchronous with hot season. The average annual temperature is 13.5℃. The average temperature of July is 27℃ while the average temperature of January is 5–7℃. The average annual humidity is 77% with 269 frostless days.

The average annual rainfall of the geopark is 1,935.6mm and can reach 2127mm in areas of rainstorm centre. The rain mostly falls in spring, summer, and autumn, and has been particularly influenced by typhoon. The drainage system is composed of mainly small and short streams due to its mountainous terrain and the water volume is significantly affected by rainfall.

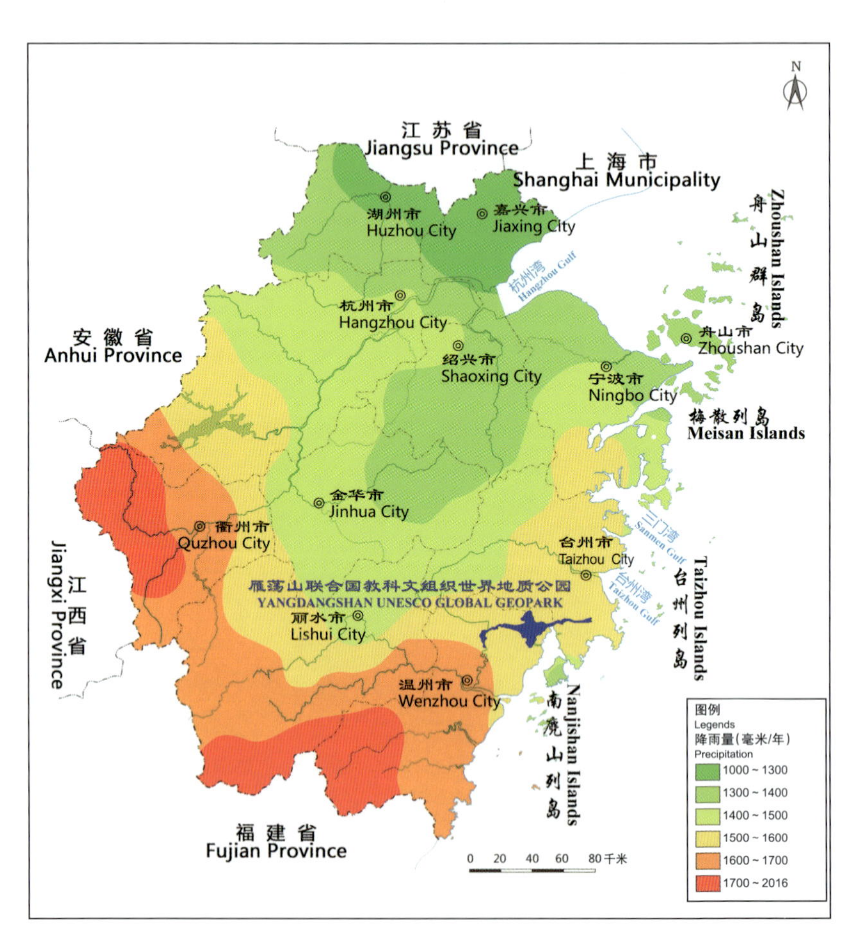

浙江省年平均降雨量图
Average Annual Rainfall of Zhejiang Province

生态特征

公园内具有极高的物种多样性，除大量海洋生物外，其他动物种类丰富，据不完全调查，区内常见的哺乳动物有8目21科53种，鸟类11目28科116种，两栖类2目8科27种，爬行动物3目9科42种，鱼类3目7科21种，昆虫12目46科221种。其中有许多珍稀濒危的野生动物，如穿山甲（*Manis pentadactyla*）、云豹（*Neofelis nebulosa*）、白鹳（*Ciconia ciconia*）、白鹤（*Grus leucogeranus*）、玉带海雕（*Haliaeetus leucoryphus*）、猕猴（*Macaca mulatta*）、大灵猫（*Viverra zibetha*）、麂（*Muntiacus*）、鬣羚（*Capricornis sumatraensis*）、白琵鹭（*Platalea leucorodia*）、小鸦鹃（*Centropus toulou*）等珍稀濒危保护动物，包括列入国家一级保护的3种，列入国家二级保护的13种。

雁荡山植物区系处于华东区系与华南区系的过渡地带，森林植被展现多样性。区内有维管束植物1659种（含引种栽培植物和种下分类群，下同），隶属189科786属，其中蕨类植物28科50属93种，裸子植物9科26属44种，被子植物152科630属1522种（双子叶植物132科510属1235种，单子叶植物20科120属287种）。此外区内发育有22种珍稀植物，大量的古树名木。特色植物有：雁荡润楠(*Machilus minutiloba* S. Lee)、雁荡山三角槭(*Acer buergerianum* Miq. var. *yentangense* Fang et Fang f.)、长叶榧(*Torreya jackii* Chun)、鹅掌楸(马褂木)[*Liriodendron chinense* (Hemsl.) Sarg.]、连香树(*Cercidiphyllum japonicum* Sieb. et Zucc.)、松叶蕨[*Psilotum nudum* (L.) Beauv.]、蛛网萼(*Platycrater arguta* Sieb. et Zucc.)、半枫荷(*Semiliquidambar cathayensis* Chang)、野大豆(*Glycine soja* Sieb. et Zucc.)、菜头肾[*Strobilanthes sarcorrhiza*]、银杏(*Ginkgo biloba*)、南方红

豹
Leopard

黑麂
Muntiacus

豆杉(*Taxus wallichiana* var. *mairei*)等。雁荡马尾杉(*Phlegmariurus yandongensis*)、金腺毛蕨(*Cyclosorus aureoglandulosus* Ching et Shing ex Chin)、秀丽野海棠(*Bredia amoena* Diels)、雁荡润楠(*Machilus minutiloba* S. Lee)等植物模式标本产地在雁荡,具有重要的科学研究价值。雁荡山古树名木共有161棵,年代比较久远的有桧柏[*Sabina chinensis* (L.) Ant.]、柏木(*Cupressus funebris* Endl.)、桂花[*Osmanthus fragrans* (Thunb.) Loureiro]、樟树[*Cinnamomum camphora* (L.) Presl]、枫香(*Liquidambar formosana* Hance)、竹柏[*Podocarpus nagi* (Thunb.) Zoll. et Mor ex Zoll.],古树名木以其古朴秀丽为特色,为雁荡山增添了无限神韵。

大灵猫
Viverra zibetha

Ecology

The geopark is rich in biodiversity with diverse species. Animal species are great in number. Surveys show that there are 53 species of mammals in 21 families in 8 orders, 116 bird species in 28 families in 11 orders and 27 amphibian species in 8 families in 2 orders. It is also the home of 42 species of reptiles in 9 families and in 3 orders, 21 species of fish in 7 families in 3 orders as well as 221 species of insects in 46 families in 12 orders. Many of which are rare and endangered wildlife, such as *Manis pentadactyla*, *Neofelis nebulosa*, *Ciconia ciconia*, *Grus leucogeranus*, *Haliaeetus leucoryphus*, *Macaca mulatta*, *Viverra zibetha*, *Muntiacus*, *Capricornis sumatraensis*, *Platalea leucorodia* and *Centropus toulou*. Three of them have been listed in first-class state protection category and thirteen of them have been listed in secondary class state protection category.

穿山甲
Manis crassicaudata

Yandangshan is located at the transitional belt of East and South China flora zones, implying great biodiversity of forest vegetation. Rare plants include vascular bundle plants (including the introduction and cultivation plants and infra-specific taxon, the same below) of 1659 species in 786 genera in 189 families, among which there are 93 species of ferns in 50 genera in 28 families. Gymnosperms include 44 species in 26 genera in 9 families. Angiosperms include 1522 species in 630 genera in 152 families (including dicotyledon of 1235 species in 510 genera in 132 families and monocotyledon of 287 species in 120 genera in 20 families). Besides, there are 22 species of rare plants as well as abundant ancient and famous woods. The characteristic plants include *Machilus minutiloba* S. Lee, *Acer buergerianum* Miq. var. *yentangense* Fang et Fang f., *Torreya jackii* Chun, *Liriodendron chinense* (Hemsl.) Sarg., *Cercidiphyllum japonicum* Sieb. et Zucc. *Psilotum nudum* (L.) Beauv., *Platycrater arguta* Sieb. et Zucc., *Semiliquidambar cathayensis* Chang, *Glycine soja* Sieb. et Zucc., *Strobilanthes sarcorrhiza*, *Ginkgo biloba* and *Taxus wallichiana* var. *mairei*. The type specimens of Phlegmariurus phlegmaria, *Phlegmariurus yandongensis*, *Cyclosorus aureoglandulosus* Ching et Shing ex Chin, *Bredia amoena* Diels and *Machilus minutiloba* S. Lee originated in Yandangshan have important scientific research value. There are 161 ancient and famous woods growing in Yandangshan. These comprise antique woods like *Sabina chinensis* (L.) Ant., *Cupressus funebris* Endl., *Osmanthus fragrans* (Thunb.) Loureiro, *Cinnamomum camphora* (L.) Presl, *Liquidambar formosana* Hance and *Podocarpus nagi* (Thunb.) Zoll. et Mor ex Zoll. These ancient and famous woods have added attraction and liveliness to the landscape of Yandangshan.

长叶榧
Torreya jackii Chun

鹅掌楸
Liriodendron chinense (Hemsl.) Sarg.

历史沿革

公园山水景观独特，环境优美。早在5000年前，新石器时代的瓯越先民就在楠溪江流域繁衍生息，造就了灿烂的瓯越文化。东晋、南宋时期，公园成为名胜之地，人们在此建寺筑庙、讲学传道。长屿硐天也在南北朝时期就开始了石材的采取，为瓯越地区提供了丰足的建筑、生产生活材料。

唐代是雁荡山的开创时期，相传西域僧诺讵那驻锡龙湫，僧贯休有“雁荡经行云漠漠”之赞，一行有“南界尽于雁荡”之语，其名始著；山中有雪洞、唐人题刻；后有僧善孜于灵峰洞（又称观音洞、罗汉洞）中颂《法华经》。宋代，雁荡山有著名的“十八名刹”。宋室南渡后，贵游辐辏，山径改辟，梵刹增新，雁荡之游始盛，诗文、题刻渐多。同期，楠溪江优美的山水风光，孕育了数百个古文化村落，成就了辉煌的“永嘉学派”。

明清时期为继续发展时期，因山水诗韵的流行，促使大量文人墨客留下大量诗文、游记、

长屿采石遗址历史发展区一角
Displaying the Quarrying History of Changyu

History

The geopark has unique landform and elegant environment. Five thousand years ago, Ou Yue ancestors in Neolithic Age lived and multiplied in Nanxijiang Basin and created the fabulous Ou Yue culture. The geopark has become a famous scenic spot since Eastern Jin and Southern Song Dynasties. It was also a selected area for temple construction, lectures and preaching. Quarrying in Changyu Dongtian started from the Northern and Southern Dynasties, providing abundant building materials for houses, infrastructure for Ou Yue regions.

Yandangshan has been developed since Tang Dynasty. It is said that Nuojunuo, a monk from the Western Regions, once lived in Longqiu and Monk Guanxiu praised the mountains and expressed that "Yandangshan touches the sea of clouds and sky". It has become famous since then and has been described as "the southern border of the country". There are Xuedong Cave in the mountains with stone inscriptions of Tang people. After that, some monks persisted in chanting the *Lotus Sutra* in Lingfeng Cave (also known as Guanyin Cave and Luohan Cave). In

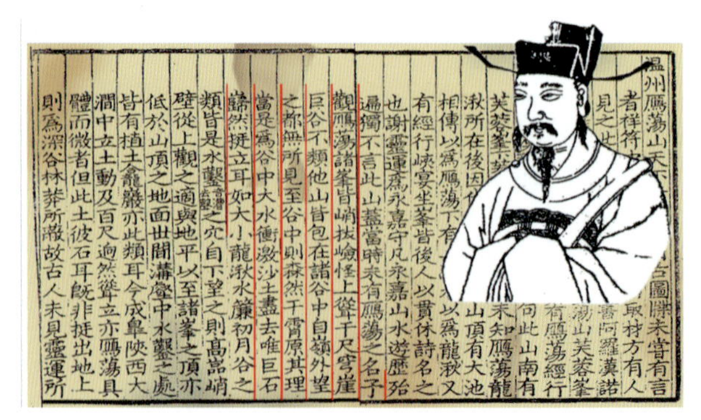

也謝靈運爲永嘉守凡永嘉山水遊歷殆
遍獨不言此山蓋當時未有鴈蕩之名予
觀鴈蕩諸峯皆峭拔嶮怪上聳千尺穹崖
巨谷不類他山皆包在諸谷中自嶺外望
之都無所見至谷中則森然干霄原其理
當是爲谷中大水衝激沙土盡去唯巨石
巋然挺立耳如大小龍湫水簾初月谷之
類皆是水鑿之穴自下望之則高嵓峭
壁從上觀之適與地平以至諸峯之頂亦
低於山頂之地面世間溝壑中水鑿之處
皆有植土龕巖亦此類耳今成皋陝西大
澗中立土動及百尺迥然聳立亦鴈蕩具
體而微者但此土彼石耳旣非挺出地上
則爲深谷林莽所蔽故古人未見靈運所

沈括与雁荡山
Shen Kuo and Yandangshan

山志、摩崖石刻，文化沉淀深厚。

1923年，公路修至白溪，内通雁荡。继而成立专门的建设管理机构，雁荡旅游掀开新的篇章。中华人民共和国成立以后，建设发展始终不辍，改革开放以来发展尤为迅速。

公园所在地1982年被国务院列为首批国家重点风景名胜区，2004年被评为国家地质公园，2005年被评为世界地质公园，2007年被评为国家首批AAAAA级旅游景区，2010年被评为国家矿山公园。

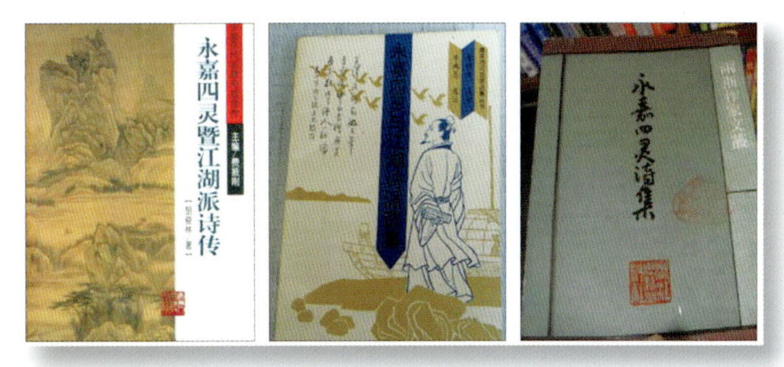

著名诗人永嘉四灵
Famous Poets: Four Poets of Yongjia

叶适与永嘉学派研究书籍
Ye Shi and Research Publications of Yongjia School of Thought

Song Dynasty, there were eighteen famous temples built. After the Song Government moved to the south nearer to Yandangshan, nobles frequently visited the areas, resulting in the increase in the numbers of mountain tracks and temples. Yandangshan peaked in history with increasing numbers of poems and stone inscriptions. Meanwhile, the elegant scenery in Nanxijiang had led to the growth of hundreds of villages with very rich local culture which eventually brought about the important "Yongjia School of Thought".

Yandangshan continued to prosper in Ming and Qing Dynasties. With scenic poems becoming popular, many poets and litterateurs wrote many poems and travel literatures, drawn mountain topographies and inscribed on precipices which completely enriched the cultural environment of Yandangshan.

In 1923, a new road was built to link up Baixi with Yandangshan. After that, a special construction and management agency had been set up and helped to open up tourism in the area. The development of Yandangshan continued after the establishment of the People's Republic of China. The growth expedites particularly after the adoption of an open approach and reform in economic policies.

The geopark was listed in the first batch of National Park of China in 1982. It was approved as National Geopark in 2004 and awarded as "Yandangshan Global Geopark" in 2005. It was also one of the members of the first batch of National AAAAA Tourist Attractions in 2007 and obtained the title of "National Mining Park" in 2010.

公园地质特征

Geological Characteristics of Geopark

区域地质背景

Regional Geology Background

中国东南大陆位于亚洲东部大陆南部边缘，在地质构造上，中国东南大陆属于环太平洋构造域，由扬子古陆块和华夏古陆块在漫长的地质演化历史中多次拼合—裂解—再拼合而成，是晚中生代火山岩-花岗岩大量分布的地区，是濒太平洋地区岩浆岩带的重要组成部分。

Southeast China is located on the southern margin of the East Asia continent. Geologically, it belongs to the circum-Pacific tectonic domain and has undergone multiple splice and split processes throughout the long geologic evolution of Yangtze-Cathaysia Plate. This is an area with extensive distribution of late Mesozoic volcanic and granitic rocks and is part of the significant magmatic belt of the Pacific Rim.

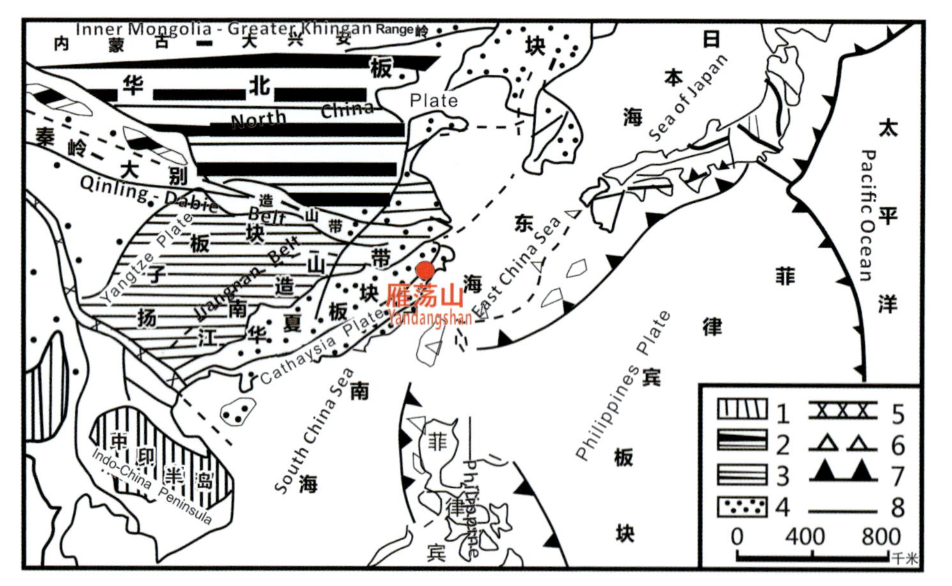

公园大地构造位置示意图
Geotectonic Location of the Geopark

地层

公园及周边出露地层自下而上可分为高坞组(K_1g)、西山头组(K_1x)、茶湾组(K_1cw)、九里坪组(K_1j)、馆头组(K_1gt)、小平田组(K_1xp)、莲花组(Qpl)和鄞江桥组(Qhy)/镇海组($Qhzh$)。

Stratigraphy

Outcrops in the geopark and surrounding areas belong to a number of different formations, starting from the oldest at the bottom to the youngest on the top: Gaowu Formation (K_1g), Xishantou Formation (K_1x), Chawan Formation (K_1cw), Jiuliping Formation (K_1j), Guantou Formation (K_1gt), Xiaopingtian Formation (K_1xp), Lianhua Formation (Qpl) and Yinjiangqiao (Qhy)/Zhenhai Formation ($Qhzh$).

年代地层 Chronostratigraphy			岩石地层 Lithostratigraphy			代号 Code	柱状图 Columnar Section	厚度 Thickness (米)	岩性 Lithology
界 Erathem	系 System	统 Series	群 Group	组 Formation	段 Member				
新生界 Cz	第四系 Q	全新统 Qh		鄞江桥组 Yinjiangqiao Formation / 镇海组 Zhenhai Formation		Qhy / $Qhzh$		2～15 / 1～75	冲积含砾砂土、亚砂土、砂砾石/海积淤泥质亚黏土、黏土、粉砂土 Alluvial gravel-bearing sand soil, sub-sandy clay and grit /Marine-deposited muddy subclay, clay, silty sand soil
		更新统 Qp		莲花组 Lianhua Formation		Qpl		5～10	冲积、坡积、洪积含砾亚砂土、砂土、亚黏土 Alluvial, slope, pluvial, gravel-bearing sub-sandy clay, sandy soil and subclay
中生界 Mz	白垩系 K	下统 Lower Series		小平田组 Xiaopingtian Formation	二段 2nd Member	K_1xp^2		＞300	流纹质熔结凝灰岩，局部夹喷溢相流纹岩 Rhyolitic welded tuff, partially interbedded with rhyolite of extrusive facies
					一段 1st Member	K_1xp^1		＞880	流纹质英安质熔结凝灰岩夹凝灰质粉砂岩 Rhyolitic dacitic welded tuff, partially interbedded with tuffaceous siltstone
			永康群 Yongkang Group	馆头组 Guantou Formation		K_1gt		453	下部为砂砾岩、砂岩、粉砂岩，上部为玄武岩、玄武安山岩夹凝灰质砂岩 Sandstone-conglomerate, sandstone and siltstone in the lower part; basalt and basaltic andesite interbedded with tuffaceous sandstone in the upper part
			磨石山群 Moshishan Group	九里坪组 Jiuliping Formation	三段 3rd Member	K_1j^3		66.5	流纹质玻屑熔结凝灰岩，局部为凝灰熔岩 Rhyolitic vitric welded tuff, partially of tuff lava
					二段 2nd Member	K_1j^2		＞1500	流纹质晶屑玻屑熔结凝灰岩 Rhyolitic crystal and vitric welded tuff
					一段 1st Member	K_1j^1		1492	流纹岩、集块角砾熔岩 Rhyolite, agglomerate breccia lava
				茶湾组 Chawan Formation		K_1cw		228	砂砾岩、凝灰质砂岩、泥质粉砂岩夹流纹质玻屑凝灰岩、沉凝灰岩 Sandstone-conglomerate, tuffaceous sandstone, argillaceous siltstone, interbedded with rhyolitic vitric tuff, sed-pyroclastic
				西山头组 Xishantou Formation	三段 3rd Member	K_1x^3		＞688	英安流纹质晶屑玻屑熔结凝灰岩、集块角砾岩 Dacite rhyolitic crystal and vitric welded tuff, agglomerate breccia
					二段 2nd Member	K_1x^2		＞582	流纹质含角砾多屑熔结凝灰岩、流纹质含角砾玻屑凝灰岩 Rhyolitic breccia-bearing poly-clastic welded tuff, rhyolitic breccia-bearing vitric welded tuff
					一段 1st Member	K_1x^1		＞1100	流纹质玻屑熔结凝灰岩、玻屑凝灰岩，夹沉凝灰岩、凝灰质砂岩 Rhyolitic vitric welded tuff, vitric tuff, interbedded with sed-pyroclastic, tuffaceous sandstone
				高坞组 Gaowu Formation		K_1g		＞1000	英安流纹质晶屑熔结凝灰岩、英安流纹质玻屑晶屑熔结凝灰岩 Dacite rhyolitic crystal welded tuff, dacite rhyolitic vitric and crystal welded tuff

雁荡山世界地质公园综合地层柱状剖面图
Composite Stratigraphic Column of Yandangshan UNESCO Global Geopark

岩浆岩

火山岩

公园内中生代火山喷发堆积了累计厚度达万米的火山喷发物，火山岩分布面积大于80%。岩石类型有凝灰岩、熔结凝灰岩、角砾凝灰岩、角砾熔结凝灰岩、集块角砾岩、集块角砾熔岩、流纹岩等。

Magmatic Rocks

Volcanic Rocks

The Mesozoic volcanic eruption in the geopark accumulated over 10, 000m of eruptive materials. And the distribution area of volcanic rocks could be more than 80%. These rocks included ash tuff, welded tuff, breccia-bearing tuff, breccia-bearing welded tuff, agglomerate breccia, lava and rhyolite.

凝灰岩
Ash Tuff

①角砾凝灰岩 Brecciat-bearing Tuff
②流纹岩 Rhyolite
③角砾熔结凝灰岩 Breccia-bearing Welded Tuff
④熔结凝灰岩 Welded Tuff

侵入岩

公园内分布的侵入岩主要为石英正长岩、正长石英斑岩，公园周边侵入岩包括碱长花岗岩、正长岩、石英二长闪长岩、二长花岗岩、二长岩、正长花岗岩、花岗岩、花岗斑岩等。除石英正长岩、正长石英斑岩、碱长花岗岩、二长花岗岩、石英霏细岩外，其他侵入岩多为岩脉。对公园及周边侵入岩采样进行Rb－Sr等时线年代测定和单颗粒锆石U－Pb法测定表明，公园侵入岩多形成于早白垩世和晚白垩世。

Intrusive Rocks

The intrusive rocks found in the geopark include quartz syenite and syenite quartz porphyry. While the intrusive rocks distributed in surrounding areas include alkali feldspar granite, syenite, quartz monzodiorite, monzonitic granite, monzonite, syenogranite, granite and granite porphyry. Except quartz syenite, syenite quartz porphyry, alkali feldspar granite, monzonitic granite and quartz felsite, most of other intrusive rocks are veins. Specimens collected in the geopark and its adjacent areas were subjected to Rb-Sr and single particles zircon U-Pb dating methods which indicated that the intrusive rocks in the area were formed during Early Cretaceous and Late Cretaceous.

百岗尖石英正长岩
Quartz Syenite in Baigangjian Mountain

构造

Structures

公园及周边分布的火山构造有雁荡山破火山、望海岗火山穹隆、章岙破火山、牛角门破火山、枫林破火山、方山破火山、晋岙破火山。

The volcanic structures distributed in the geopark and surrounding areas include the Yandangshan Caldera, Wanghaigang Volcano Dome, Zhang'ao Caldera, Niujiaomen Caldera, Fenglin Caldera, Fangshan Caldera, and Jin'ao Caldera.

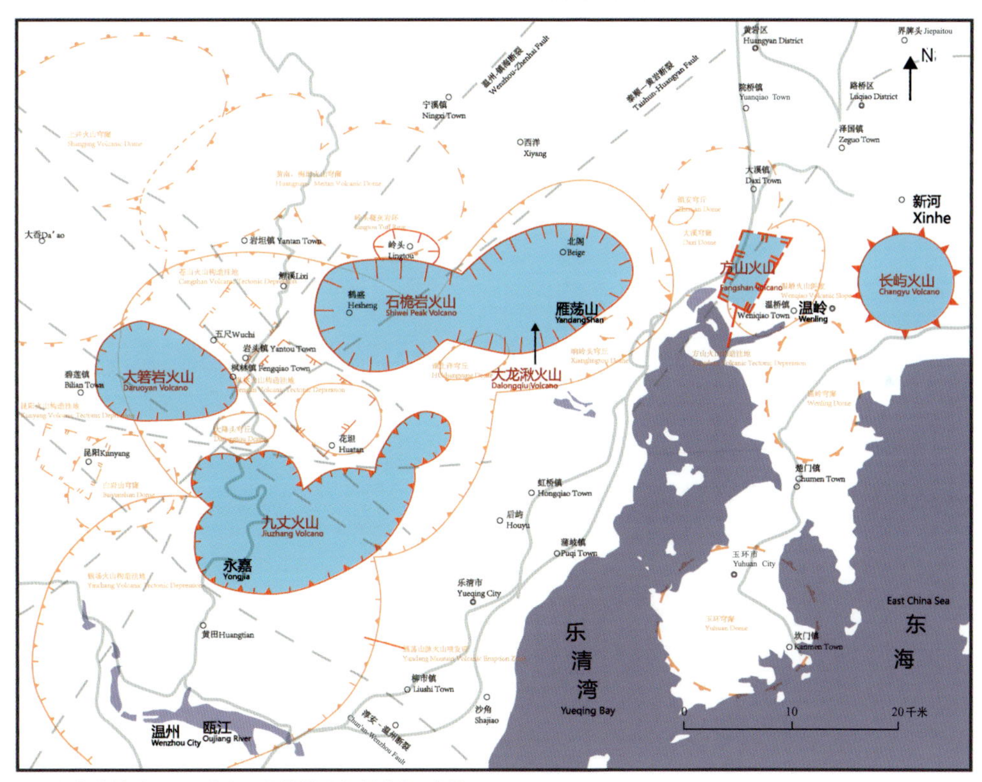

1.15亿年前楠溪江—雁荡山—长屿地区火山活动示意图
Volcanic Activities of Nanxijiang-Yandangshan-Changyu Region 115 Million Years Ago

火山活动期次划分

公园经历了4期火山喷发、2次破火山口塌陷与复活、1期岩浆侵入过程。雁荡山破火山形成演化的历史总结为7个阶段：

（1）第一期喷发阶段：大规模普林尼式爆发，喷出厚层火山碎屑流相低硅流纹质熔结凝灰岩（K_1x^3），对应岩石主要分布在雁湖岗周边、方山外围、长屿地区及楠溪江沿岸。

Stages Division of Volcanic Activities

The geopark experienced four stages of volcanic eruptions, two stages of caldera formation and revival and one major magma intrusion. Therefore, the evolution of Yandangshan caldera had gone through a total of seven stages of development:

(1) Eruption (Stage Ⅰ): Thick layer of pyroclastic fluid phase low-silicon rhyolitic welded tuff (K_1x^3) was erupted from massive Plinian eruption. The corresponding rocks distributed mainly in surrounding area of Yanhugang Mountain, periphery of Fangshan, Changyu Region and Nanxijiang.

雁荡山火山第一期爆发形成火山碎屑流
The 1st Stage of Volcanic Eruption in Yandangshan Triggered Pyroclastic Flow

(2)第一次破火山形成阶段:大规模火山爆发之后,岩浆房排空,引发火山口塌陷,导致岩层产状围斜内倾,并形成环状和放射状断裂,岩浆沿断裂侵位,形成侵出相英安流纹质凝灰熔岩岩穹。

(2) The first formation of caldera: After a massive volcanic eruption, the magma chamber was emptied and collapsed, leading to the formation of caldera, inclined strata towards the crater, radiated joint and fault patterns, magma intrusion along faults and the formation of extrusive dacitic rhyolitic tuffaceous lava dome.

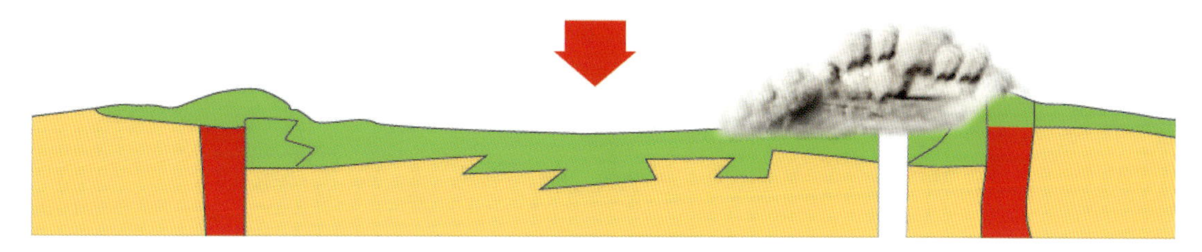

破火山塌陷,局部蒸汽岩浆爆发
The Caldera Collapsed and Phreatomagmatic Eruption Happened in Some Areas

(3)第二期喷发阶段:为早期破火山复活阶段,呈大规模喷溢流式爆发,喷出溢流相厚层流纹岩,局部普林尼式爆发,形成火山碎屑流相熔结凝灰岩,晚期还有侵出相流纹斑岩岩穹(K_1xp^1下部),对应岩石主要分布于雁湖岗周边。

(3) Eruption (Stage Ⅱ): It was the early regrowth stage of the caldera characterised by massive effusion to form thick rhyolite layers. Some areas had experienced Plinian eruption to form pyroclastic flow forming welded tuffs and porphyritic rhyolite dome (K_1xp^1 lower part) in later period. These rocks are extensively found around Yanhugang Mountain.

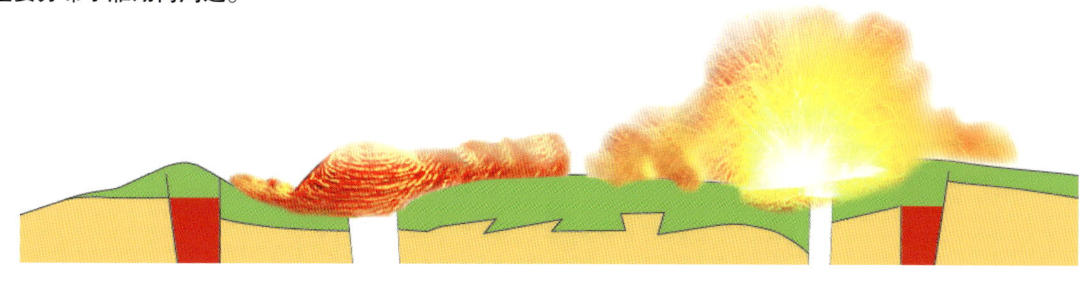

雁荡山火山第二期,破火山复活,火山喷溢,熔岩溢流与侵出
Revival of Caldera During the 2nd Stage of Volcanic Eruption Followed by Extrusion of Lava to the Surrounding Areas

(4)第三期喷发阶段：小规模亚普林尼式喷发，产物为火山碎屑流相熔结凝灰岩、凝灰岩，局部夹有溢流相流纹岩(K_1xp^1上部)，对应岩石主要分布于雁湖岗周边。

(4) Eruption (Stage Ⅲ): It was a small scale Plinian eruption which formed pyroclastic welded tuff and tuff with partially effusive rhyolite (K_1xp^1 upper part). These rocks were mainly located around Yanhugang Mountain.

雁荡山火山第三期，火山再次局部爆发，形成空落、火山碎屑流
In the 3rd Stage of Volcanic Eruption, the Volcano Produced Ashfall and Pyroclastic flows

(5)第四期喷发阶段：为雁荡山最后一次全区性普林尼式猛烈爆发，形成火山碎屑流相流纹质熔结凝灰岩，局部有溢流相凝灰熔岩(K_1xp^2)，对应岩石主要分布于雁湖岗周边。

(5) Eruption (Stage Ⅳ): This was the last but violent Plinian eruption which affected the whole region and formed rhyolitic welded tuff by pyroclastic flow. It also formed effusive tuffaceous lava (K_1xp^2). The relevant rocks are mainly found around Yanhugang Mountain.

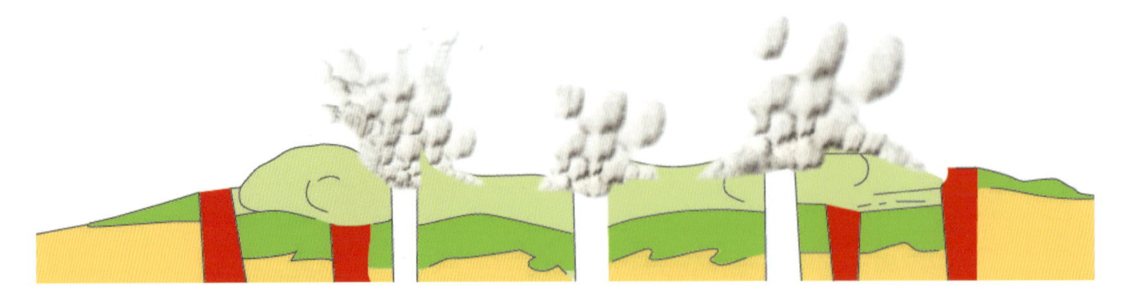

雁荡山火山第四期，破火山再次爆发形成火山碎屑流
The 4th Stage of Volcanic Eruption in Yandangshan Involved Extensive Pyroclastic Flow

(6)第二次破火山形成阶段:在第二到第四次大体积岩浆爆发之后,岩浆房再次排空,火山口再次发生塌陷,导致火山岩层产状围斜内倾。

(6) The second formation of caldera: After the second to fourth massive magma eruption, the magma chambers were emptied and its roof collapsed to the crater producing periclinal-inner inclination of the volcano strata.

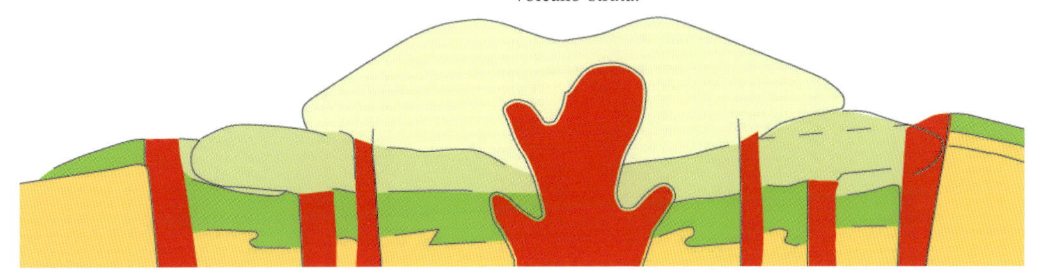

破火山再次塌陷,岩浆侵入
The Caldera Collapsed Again and Followed by Magma Intrusion

(7)中央侵入体形成阶段:晚期破火山复活阶段,岩浆沿主要喷发中心上升侵位形成中央侵入相石英正长岩($K_2\xi o\pi$),主要分布于雁湖岗及石桅岩周边。后期有英安玢岩、流纹斑岩、斜长霏细斑岩、霏细斑岩等酸性岩墙和岩脉沿破火山环形与放射状断裂侵入。

(7) Central intrusion formation: This happened in the later period of the revival of the caldera. Magma extruded according to the eruption centre to form the central intrusive quartz-syenite ($K_2\xi o\pi$), which could be found around Yanhugang and Shiwei Peak areas. At the later stage, the intrusion of acid dykes and veins such as dacite porphyrite, rhyolite porphyry, plagioclase felsite porphyry and felsite porphyry predominated along the ring and radiated fractures.

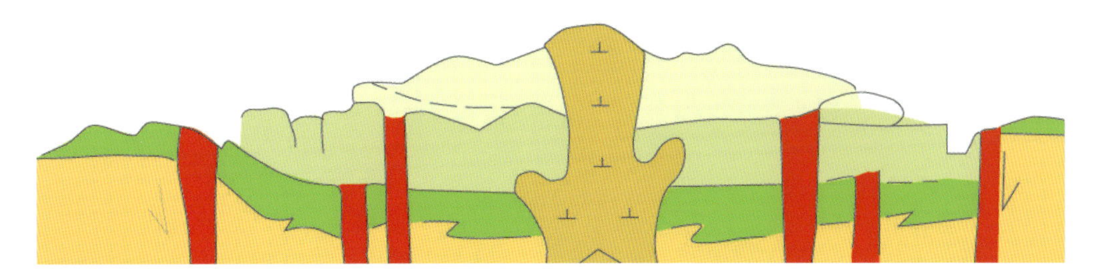

破火山抬升与剥蚀
Uplift of Caldera Enhanced Erosion

楠溪江园区处在雁荡山破火山边缘，火山喷发活动时序与雁荡山园区一致，在空间上相连，也形成与主中心一致的4期喷发、1期侵入过程，但喷发间断在楠溪江园区有很好的反映。3期间断均形成水下沉积，更好地表明沉积间断的时间以及间断时期的外动力地质过程，这对于更准确研究火山喷发循环有重要的价值。

Nanxijiang Scenic District is located at the edge of the Yandangshan caldera. The time of its volcanic eruption is the same as in the Yandangshan Scenic District—they were well-connected and have undergone four stages of eruption and one intrusion in conformity with the main centre. But there are evidences of intermittent eruption in Nanxijiang Scenic District. Underwater deposits were formed during the three eruption intervals. They demonstrated deposition and the exogenic geological processes during these intervals and had played a significant role in accurate research on the eruption cycle.

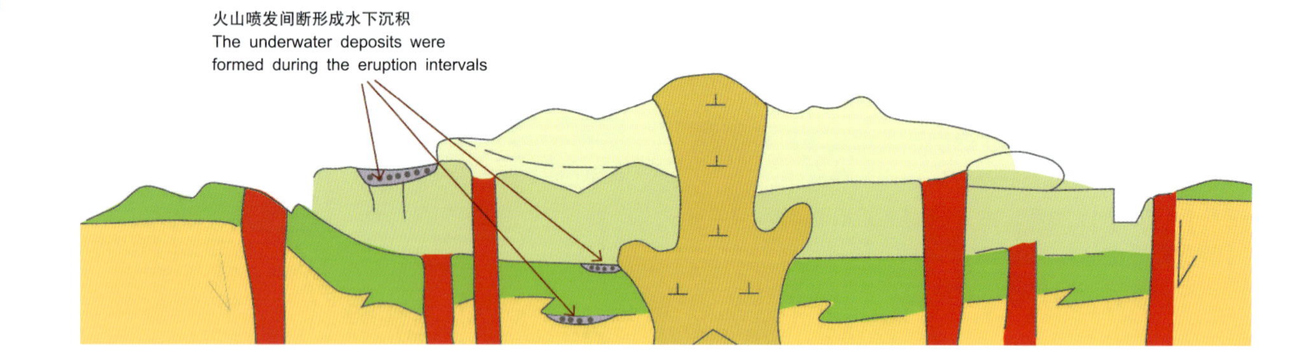

Scientific Value

雁荡山世界地质公园地处环太平洋亚洲大陆边缘的南部——中国濒太平洋构造域。在中生代时期，隶属古太平洋板块的库拉板块向亚洲大陆板块低角度斜向俯冲，导致中国东部由东西向构造体制转变为北东—北东东向构造体制。这一构造运动相当于中国的燕山运动。地处中国东南沿海的雁荡山白垩纪破火山，即在这一具有全球意义的构造背景下形成，因而雁荡山白垩纪破火山在古火山学、火山岩石-岩相学、区域构造学、大地构造学和地貌学上均具有重要的科学价值。

Yandangshan UNESCO Global Geopark is located at the south of the Asian continent margin, the marginal-Pacific tectonic domain of China. In the Mesozoic Era, a low angle, oblique subduction occurred by the Kula Plate (a part of the paleo-Pacific plate) to the Asian plate, which caused a transformation of the EW tectonic system of east China to the NE-NEE tectonic system. This tectonic movement was similar to the Yanshan movement of China. Yandangshan Cretaceous caldera is located in the Chinese southeast coastal areas. The tectonic characteristics have important global significance and scientific values in paleo-volcanology, volcanic petrography, regional geo-tectonics and geomorphology.

古火山学意义

雁荡山地质遗迹堪称中生代晚期亚洲大陆边缘复活型破火山形成与演化模式的典型范例。它记录了火山爆发、塌陷、复活隆起的完整地质演化过程，为人类留下了研究中生代破火山的一部永久性文献，成为一个天然的破火山立体模型和研究白垩纪破火山的野外实验室。

雁荡山破火山形成后，在将近1亿年的地质历史时期内，地壳总体处于抬升。第四纪高海平面时，海水仅到达其外缘沟谷。破火山变形微弱，以沿断裂切割、抬升为主，使火山岩层和侵入体——石英正长岩露出地表。在环状、放射状断裂或区域性断裂的基础上发育沟谷，切割出了破火山内部的岩石层序和构造断面。可以说，是大自然的力量解剖了这座破火山，使之成为一个天然的立体模型，向人们清楚地展示出白垩纪破火山的内部结构，特别是破火山根部带的地质构造要素和各类岩石的相互关系，出露了破火山的喷发产物与侵入体的相互关系，记录了早期酸性岩浆演化及侵入冷却结晶的全过程，这对于研究白垩纪破火山的形成及其岩浆作用有着重要的科学价值。

Paleo-Volcanology Significance

The geoheritage of Yandangshan is a formation and evolution representation of revived caldera in the Asian continent margin of the late Mesozoic Era. It recorded the complete geological evolution process, including eruption, collapse, regrowth and uplifting of the volcanoes in Yandangshan, forming an ideal 3D model, a field study and outdoor research laboratory for Cretaceous calderas.

Uplifting followed the formation of the Yandangshan Caldera 100 million years ago. When sea level rose in Quaternary, the water only reached to its outer valleys. The caldera transformed mainly by cutting and uplifting along fractures. Quartz syenites were formed by intrusion and later appeared on ground due to erosion. Valleys developed around the ring and along radiated fractures. Lithologic sequences and tectonic sections of the caldera were therefore exposed by the power of nature, revealing the internal structure of the Cretaceous caldera, especially its relationship between the caldera root and tectonic elements and rocks of the caldera root. It also revealed the interaction between eruption products and the intrusion, recording the whole process from early acid magma evolution and intrusion to cooling and crystallization. The record has important scientific value for studying the formation and magmatism of Cretaceous calderas.

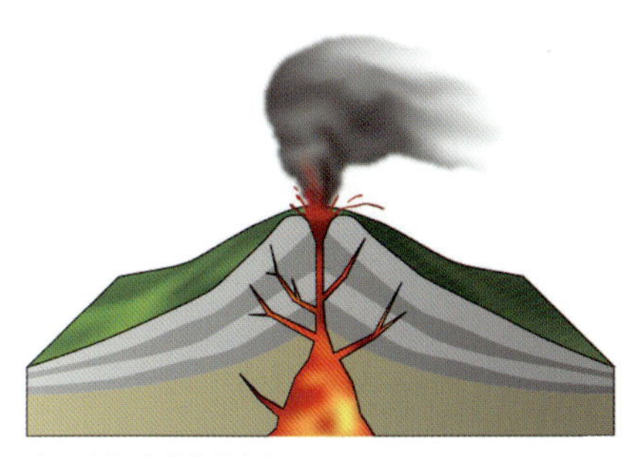

火山喷发，岩浆大量喷出
Magmatic Eruption

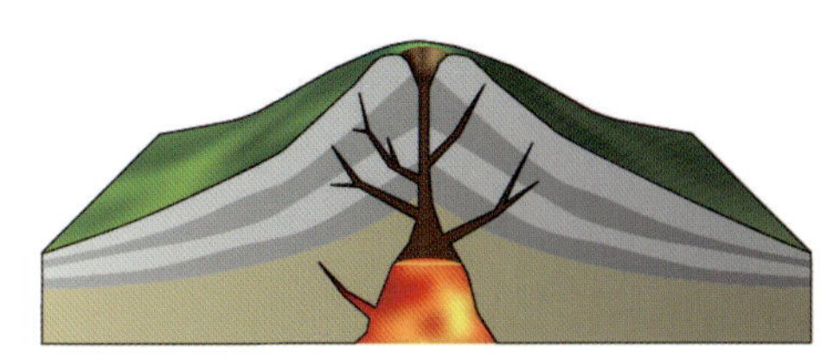

岩浆房排空
Emptied Magma Chamber

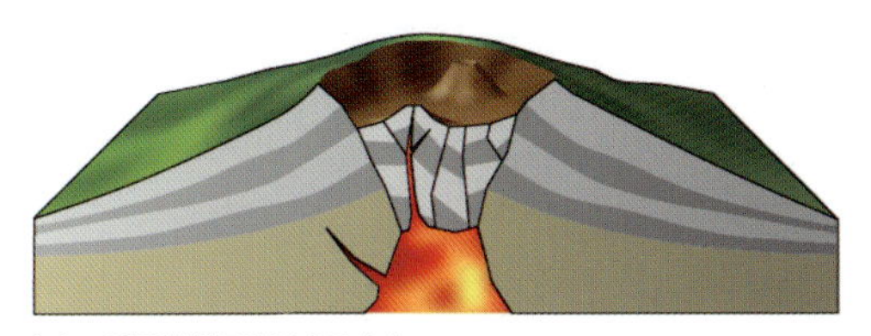

火山口周围崩塌下陷形成破火山
The Collapse and Subsidence of Crater to Form Caldera

破火山形成图解
Evolution of Yandangshan Caldera

火山岩石-岩相学意义

Volcanic Rocks and Petrology Significance

雁荡山破火山是酸性岩浆经爆发、喷溢、侵出及侵入形成的。其产物涵盖了不同岩相的岩石，包括地面涌流堆积、火山碎屑流堆积、空落堆积、基底涌流堆积和流纹质熔岩、岩穹、次火山岩等。岩石地层单元、岩相剖面、岩流单元及岩石结构均十分典型，它几乎包括了岩石学专著中所描述的流纹岩类各种岩石。因而，雁荡山被称为流纹质火山岩天然博物馆。

Yandangshan Caldera was a combined result of acidic lava explosion, effusion, extrusion and intrusion forming different rock facies, accumulation of ground surge, pyroclastic flow, air fall, base surge, rhyolitic lava, dome and secondary volcanic rocks. It also displays vivid lithological units, profiles, flow units and structures. Yandangshan has almost all types of rhyolitic rocks and is therefore named the Natural Museum of the Rhyolitic Volcanic Rocks.

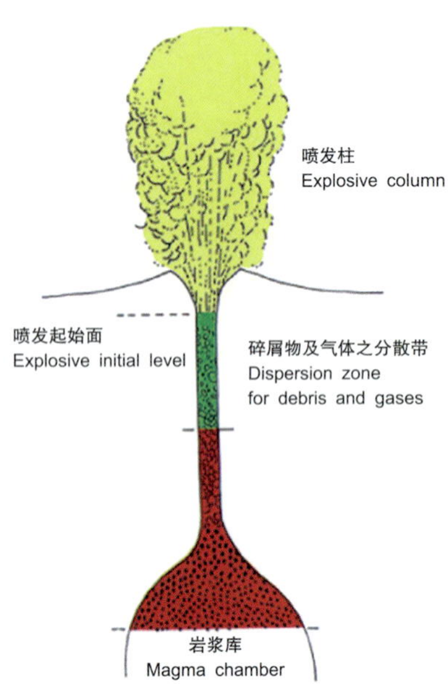

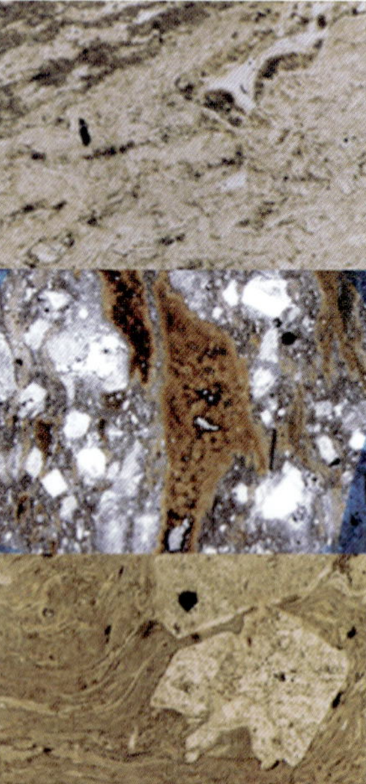

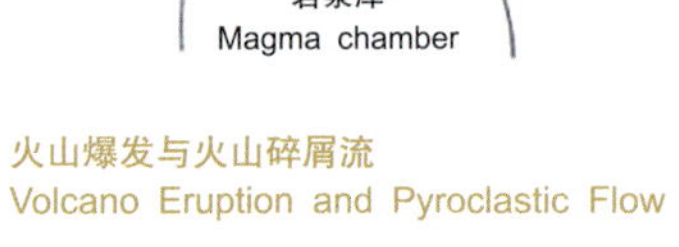

火山爆发与火山碎屑流
Volcano Eruption and Pyroclastic Flow

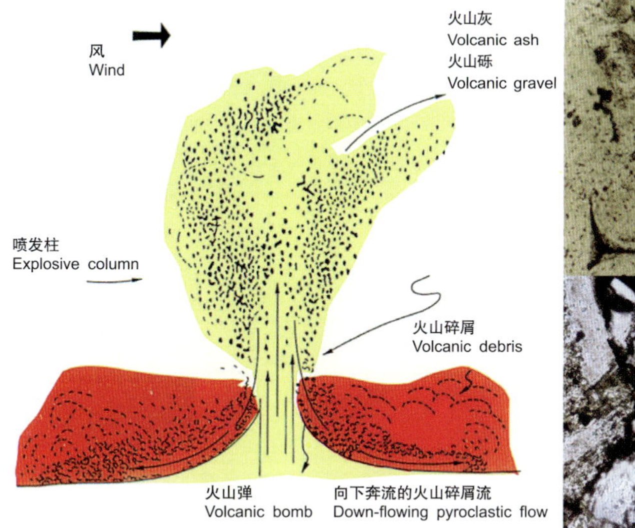

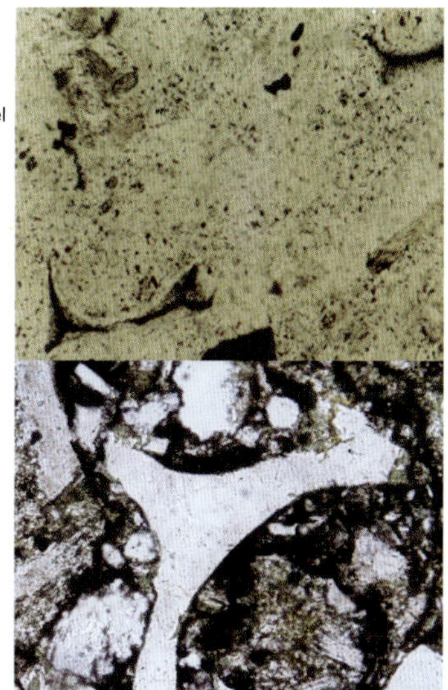

火山爆发与空落堆积
Volcano Eruption and Airfall Accumulation

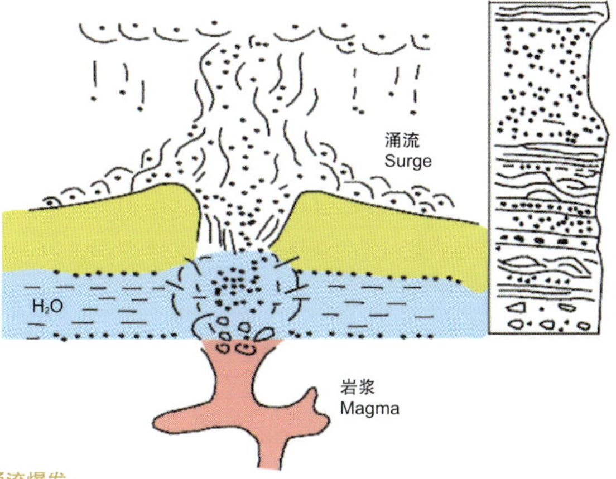

基底涌流爆发
Base Surge Eruption

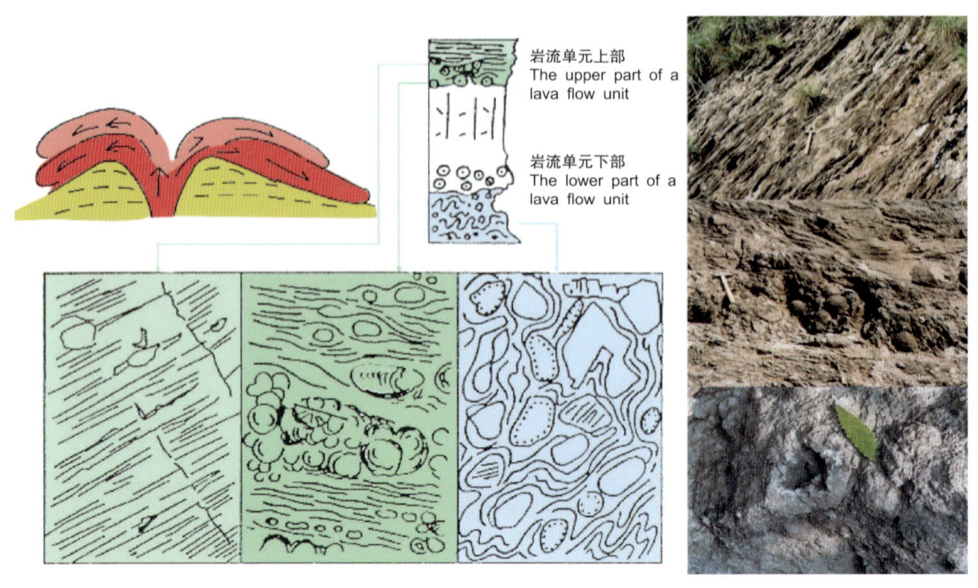

火山岩浆从火山口中沿地表溢流形成岩流单元
Lava Extruded from Vent and Flowed Slowly on Ground to Form Flow Units

岩穹形成示意图
Formation of Lava Dome

区域地质构造意义

雁荡山破火山是中国燕山运动中惊天动地岩浆大爆发的一个典型代表。中国东南沿海火山岩带形成于燕山期，呈北东—北北东向展布。该岩带以酸性岩浆的爆发占主导地位，大面积的火山碎屑流堆积，包括未熔结、熔结凝灰岩，其岩浆爆发的体积约48万立方千米。其规模可与俄罗斯东锡霍特-阿林、澳大利亚东北部、美国西部流纹质熔结凝灰岩相比。雁荡山破火山4期火山活动中有3期为火山爆发，可作为燕山期火山岩浆大爆发的一个典型区，这对于研究中国燕山期构造-岩浆活动及其在东南沿海表现形式具有重要的科学价值。

Regional Geological Structure Significance

Yandangshan Caldera is a typical representative of the earth-shattering magma eruption of Yanshan Movement of China. Chinese southeast coastal volcanic belt was formed during the Yanshan Movement in a NE−NNE- trending direction. Acid lava dominated in this volcanic belt in form of extensive volcaniclastic flow accumulation, including tuff and welded tuff. The volume of the eruption is around 480, 000km^3 comparable to the rhyolitic welded tuff in Russia Sikhote-Alin, northeast Australia and western United States. There are three volcanic eruptions in the four stages of volcanic activities of Yandangshan and has been regarded as a typical magma explosion of Yanshan Period. This is of significant scientific value particularly in the study of the magmatic activities associated with the Yanshan tectonic movement along the south eastern coast of China.

全球地质历史演化对比意义

雁荡山白垩纪流纹质破火山在中生代西太平洋大陆边缘火山带演化中具有典型性。西太平洋亚洲大陆边缘为巨型中生代火山岩带，以发育流纹质火山岩为特色。其分布范围北起俄罗斯的鄂霍茨克-楚科奇、东锡霍特-阿林，经朝鲜半岛，南至中国东南部；东太平洋北美大陆边缘则以发育安山岩为特色，主要分布在墨西哥、秘鲁、智利等地。

在中生代时期，隶属于古太平洋板块的库拉板块向亚洲大陆板块斜向低角度俯冲，玄武质岩浆底侵作用提供热源，诱发地壳中岩浆岩、变质岩部分熔融产生高钾钙碱性岩浆。这种酸性岩浆沿北东—北北东向断裂喷发，在地表构成了如今中国东南沿海火山岩带。

雁荡山白垩纪破火山为中国东南部火山岩带的代表。从这一意义上讲，雁荡山破火山的形成与中生代全球性板块运动有关。

雁荡山世界地质公园的地质遗迹蕴含着中生代时期伊泽奈基-库拉-太平洋板块与亚洲大陆板块相互作用及其深部岩浆作用过程的重要信息，为研究亚洲大陆边缘动力学提供了火山学与岩石学证据。

Global Geological Evolution & Comparison Significance

The Cretaceous rhyolitic caldera of Yandangshan is typical in the evolution of the volcanic belt of the west Pacific continental margin in the Mesozoic Era. The continental margin of west Pacific is a huge Mesozoic volcanic rock belt, which features the occurrence of rhyolites. The belt extends from Russia Okhotsk Chukchi and East Sikhote-Alin, then through southward to the Korean Peninsula and further southward to the southeastern part of China. On the other side of the Pacific Ocean, however, andesites dominate the North American continental margin and extend south to Mexico, Peru and Chile.

In the Mesozoic Era, a low angle, oblique subduction occurred by the Kula plate (a part of the Paleo-Pacific plate) to the Asian plate. Basaltic magma provided heat source for underplating and caused melting of the magmatic rocks and metamorphic rocks which generated high-K calc-alkaline magma. This acid magma erupted along the NE–NNE-trending faults, bringing into being the coastal volcanic rock belt on the surface in southeast China.

Yandangshan Cretaceous caldera is an excellent representation of the volcanic rock belt in southeast China. Therefore, the formation of Yandangshan caldera is closely associated with the global plate movement during the Mesozoic Era.

The geological heritages of Yandangshan UNESCO Global Geopark provide important information of the interactions between the Mesozoic Izanagi-Kula-Pacific plate and the Asian plate as well as the their deep magmatic processes, substantiating volcanological and petrological evidences for studying geo-dynamics of the Asian continental margin.

雁荡山火山与古太平洋板块俯冲
Volcanoes of Yandangshan and Subduction of Paleo-Pacific Plate

人文景观

雁荡山世界地质公园是自然与人文景观的完美组合，是“天人合一”的伟大创造。这里丰富的人文景观渗透在雄奇秀美的自然山水之中，构成了旅游资源的重要组成部分。公园内遍布崖岩碑刻、寺院道观、纪念馆、古亭、古桥、古塔、古墓等。自唐宋以来描述点评雁荡山地区的诗词达5000多首，志书游记30多部，集文学与书法、石刻艺术于一体的摩崖石刻400多处。永嘉昆剧更是被评为世界级非物质文化遗产。如今，当人们走进这个地质公园时，不仅可以领略到雄伟壮丽的火山岩地貌景观、优雅恬静的楠溪江，还能欣赏古色古香的村落、恢宏奇特的长屿硐天古采矿遗迹和渊源流长的石文化。

Cultural Landscapes

Yandangshan UNESCO Global Geopark is a perfect combination of natural and cultural landscapes and a representation of "unification of humans and nature". Such abundant cultural landscapes enhance its tourist attraction to become invaluable tourism resources with the presence of numerous rock carvings and inscriptions, Buddhist and Taoist temples, memorials, historical pavilions, bridges, towers and tombs. There are over 5,000 famous poems, 30 books, 400 stone sculptures having been described and commented on Yandangshan since Tang and Song Dynasties. The Yongjia Kun Opera has been listed as world's intangible cultural heritage. When visiting the geopark, in addition to the fascinating volcanic landform and the Nanxijiang, the history and culture of the area such as the quarry heritage of Changyu Dongtian and the longstanding culture of stones are also very appealing and attractive.

硐天花园
Dongtian Garden

摩崖石刻
Cliff Rock Carving

岩头古村永嘉昆剧表演
Yongjia Kun Opera Performance in Yantou Ancient Village

科考价值

雁荡山地质地貌启智功能早在11世纪就对我国科学家产生作用，是沈括流水侵蚀学说思想的发源地。近代中国地质学家基于对中国东部中生代火山地质与岩石学研究，认为雁荡山为白垩纪破火山与流纹质火山岩的典型地区，已成为我国这一学科的教学基地。自1996年被第三十届国际地质大会专家考察以来，雁荡山已引起国内外专家的广泛关注，成为考察研究中国东部火山岩带的经典地区。

Value in Scientific Research

The geology and geomorphology of Yandangshan has stimulated the interest of early Chinese scientists since 11th century. It is the origin of the great scientist Shen Kuo's water erosion theory. It is also a favourable research venue for Chinese geologists to study Mesozoic volcanic geology and lithology in East China. It is a representation of typical Cretaceous calderas and rhyolitic rocks. Since the field trip conducted by experts from the 30th International Geological Congress in 1996, it continues to be visited by many domestic and international experts to study the volcanic belt in East China.

考察路线

Field Trip Routes

流纹质火山岩地貌考察路线

Field Trip Routes for Rhyolitic Volcanic Rock Landform

博物馆-灵峰景区

Museum－Lingfeng Scenic Area

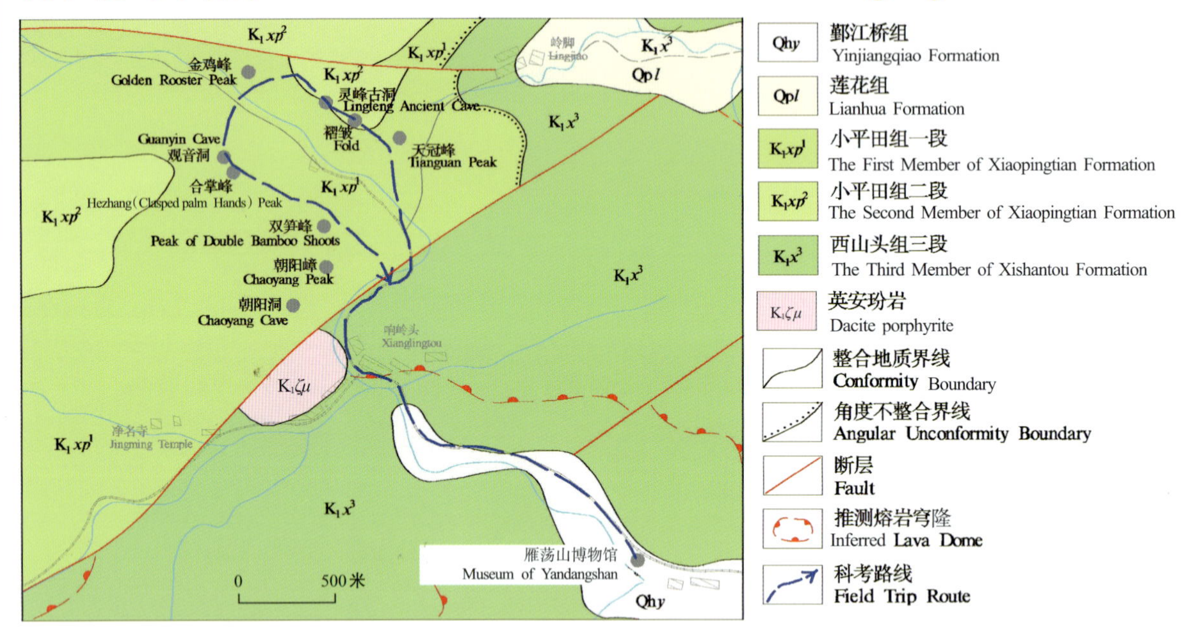

博物馆—灵峰景区考察路线图
Route for Field Trip of Museum－Lingfeng Scenic Area

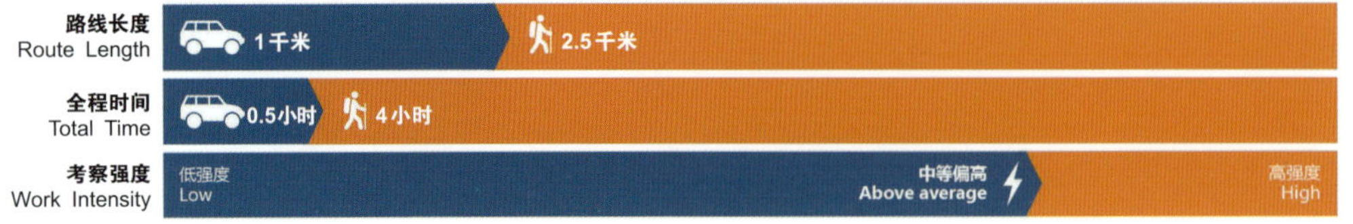

考察强度中等偏高，东西瑶台、观音洞、北斗洞栈道较陡

The work intensity is above average. The plank roads of the East Yaotai, West Yaotai, Guanyin Cave and Beidou Cave are rather steep.

博物馆
Museum

目的 Purpose	参观雁荡山博物馆 To visit Museum of Yandangshan
位置 Location	雁荡镇西北约1千米 About 1km northwest of Yandang Town
路线 Route	博物馆室内通道 Museum's indoor route

雁荡山博物馆占地面积15 333平方米，建筑面积3620平方米。分8个展厅和3个展廊，即影视厅、序厅、地质遗迹厅、火山演示厅、文化厅、世界火山厅、生态厅、地质公园建设发展厅，地质展廊、科普展廊及姊妹公园展廊。馆内充分展示了雁荡山地质地貌特征、地质遗迹特征、全球火山特征、火山喷发和破火山形成过程、生物多样性以及地质公园发展历程。博物馆建成至今已接待了多位国家领导人、国内外专家及大量游客，也成为了青少年科学考察探险基地、国土资源科普基地、全国科普教育基地。

Museum of Yandangshan has eight exhibition halls and three exhibition galleries, i.e., cinema, preface hall, geoheritage hall, volcano display hall, culture hall, world volcano hall, ecology hall, geopark development hall, geology gallery, science popularization gallery, and geopark gallery. The geological and geomorphologic characteristics, geoheritage properties, volcano eruption and caldera formation processes, biological diversity, development history of the Geopark and the worldwide volcanos are well displayed. The museum has entertained many national leaders, experts and a large number of visitors at home and abroad since its establishment and also has become the Adolescent Scientific Expedition and Exploration Base, the Land and Resources Science Popularization Base, and the National Science Education Base.

灵峰景区
Lingfeng Scenic Area

目的 Purpose	考察灵峰景区独特的流纹质火山岩地貌及人文特色 To investigate the unique rhyolitic volcanic rock landform as well as the cultural characteristics in Lingfeng Scenic Area
位置 Location	雁荡镇西北约3.5千米 About 3.5km northwest of Yandang Town
路线 Route	灵峰景区栈道 Plank road of Lingfeng Scenic Area

灵峰景区是雁荡三绝“二灵一龙”之一，以1亿年前火山溢流喷发形成的流纹岩，并经后期风化形成的悬崖叠嶂、奇峰怪石、幽深石洞、清润碧潭而著称。景区内奇峰环绕，千形万状，美不胜收；畅游夜景，移步换形，变换多姿，昼夜不同，妙不可言；更有观音奇洞，高筑九层楼阁，堪为雁荡第一洞天。

Lingfeng Scenic Area is one of the three most popular sites in Yandangshan. The rhyolitic landscape is formed by effusive volcanic eruptions about 100 million years ago. Weathering of the volcanic rocks creates spectacular peaks, cliffs and caves. The scenery is changeable when being viewed from different angles and greatly varies during daytime and at night. The Guanyin Cave together with the nine storey tall pavilions is the most fascinating cave in Yandangshan UNESCO Global Geopark.

朝阳嶂

朝阳嶂横400余米、高100余米，展如屏，面向东，故名。它是由雁荡山第二期火山喷发溢流形成的流纹岩层构成，受断裂影响，经风化和流水作用形成了如今的嶂壁。

Chaoyang Cliff

Chaoyang Cliff is an east-facing, screen-like rock surface of 400 metres long and 100 metres high. It is shaped by faulting, weathering and water erosion on the rhyolitic layers formed during the second stage of volcanic effusion of the Yandangshan Caldera.

叠嶂形成示意图
Formation of the Cliffs

双笋峰

两峰矗立，高80余米，其形如双笋并立。这是由于流纹岩层中发育两个方向的裂隙，在裂隙处岩石较破碎，后经风化作用和流水侵蚀，岩石崩塌，残留下两个峰柱。

Peak of Double Bamboo Shoots

These two peaks have a height of over 80 meters with the shape of two bamboo shoots standing upright and pointing to the sky. They are the results of weathering and water erosion along two joints (or fractures) running in two different directions. The rocks along the joints were therefore weakened and later collapsed to separate the rhyolite block into two distinctive sharp peaks.

双笋峰
Peak of Double Bamboo Shoots

天冠峰

天冠峰峰高120余米，宽150余米，状如礼帽。其岩石是雁荡山火山距今1.21亿年喷溢的熔岩——流纹岩。其上部保留了熔岩流动的痕迹——近水平的流纹构造；下部为含角砾球泡流纹岩，易剥落成小型洞穴，峰下的响板岩洞即为一例。

Tianguan Peak

With a height of more than 120 meters and a width of more than 150m, the peak looks like a hat. The rocks are rhyolites formed by lava effusion of Yandangshan volcano 121 million years ago. There are evidences of flow-banded structures which are retained in its upper part——nearly horizontal rhyolitic structures. The lower parts are breccia-bearing lithophysal rhyolite, which can be easily exfoliated to form small caves. Xiangbanyan Cave under the peak is one of the good examples.

天冠峰
Tianguan Peak

流动构造

火山碎屑涌流属于地面涌流，火山喷发产物形成热气底浪携带着细粒火山碎屑沿古地貌运移前进并沉积下来。火山碎屑涌流堆积的特征之一为分布范围小，厚度薄，特别是古地貌起伏不平时具有填平补齐作用。当时的古地貌蜿蜒曲折决定了火山碎屑流凝固成岩石后呈现出现在的形态，这种现象在地质学上叫作流动构造。

Flow Structure

Volanic clastic surge is a ground surface surge. The waves containing hot gas generated by volcanic eruption products, such as the fine-tephra, moved along the ancient landform and settled. The surge was characterised by its limited coverage with thin layers, filling up and levelling the irregularity of ancient land surface. The current form of surge tuff was determined by the topography of the original landscape and is called flow structure.

流动构造
Flow Structure

灵峰古洞

灵峰古洞位于金鸡峰下，鸣玉溪畔，俗称倒灵峰。天圣元年（公元1023年）曾建寺，在600多年前因断裂构造，发生山崩地裂，岩块向下滑移、崩落，在山根处架空堆积成洞穴群。这是发生在元代的一次地质灾害形成的遗迹。洞穴形态各异，洞洞相连，迂回曲折，深幽奇特。现有云雾、透天、含珠、隐虎、好运、玲珑、凉风7洞。

Lingfeng Ancient Cave

Located under Golden Rooster Peak, with Mingyu Stream by the side, Lingfeng Ancient Cave is also called Daoling Peak. There was once a temple built in the first year of Tiansheng Era（A.D.1023）, which was destroyed 600 years ago by landslide. The fallen rocks accumulated at the foothill of the mountain form the caves and arches. It is a geological heritage of disaster nature in Yuan Dynasty. The caves are strange in various shapes, linked with each other by winding paths. There are a total of seven caves, namely Yunwu, Toutian, Hanzhu, Yinhu, Haoyun, Linglong and Liangfeng.

灵峰古洞
Lingfeng Ancient Cave

合掌峰、观音洞

合掌峰是雁荡山代表景观之一，峰高约270米，在群峰环拱中直插云霄。该峰岩石为火山喷溢的流纹岩，后因断裂作用将一峰开裂为如合掌之两峰，左称灵峰，右称倚天峰，两峰之间为天然洞府。

观音洞位于合掌峰内，洞穴是由于垂直裂隙发育，岩石崩落而成。洞高113米，宽14米，深76米。倚洞就势而建九层楼阁。建筑与洞穴完美结合，巧夺天工。国内外人士倍加赞赏，邓拓诗曰“两峰合掌即仙乡，九叠危楼洞里藏”。洞内有“洗心”“漱玉”“石釜”三泉。洞内阁楼是观景最佳处。

Hezhang (Clasped Palm) Peak, Guanyin Cave

Hezhang Peak has a height of 270m and is one of the most important icons of Yandangshan. It is a rhyolite peak being cut in the middle by a fault to form the shape of two hands. The left one is called Lingfeng and the right one is called Yitianfeng. A natural cave was formed between them.

Guanyin Cave is located inside the Hezhang Peak. The cave is developed along vertical joints associated with the small faults. It has a height of 113m and is 14m wide and 76m deep. By taking advantage of the form of the cave, a nine-storey buddhist temple was built. The building is magnificent which fuses naturally with the geology of the peak. It is a favourite attraction for all visitors. Deng Tuo, a modern Chinese poet, described the scene in his poem that "Two peaks being put together to form the Hezhang Peak (a Buddhist fairyland) with the hidden nine-storey Buddhist temple inside the Guanyin Cave". There are three springs in the cave, namely Xixin, Shuyu and Shifu. The penthouse of the temple has the best view of the surrounding scenery.

合掌峰、观音洞
Hezhang Peak, Guanyin Cave

灵岩-方洞景区

Lingyan—Fangdong Scenic Area

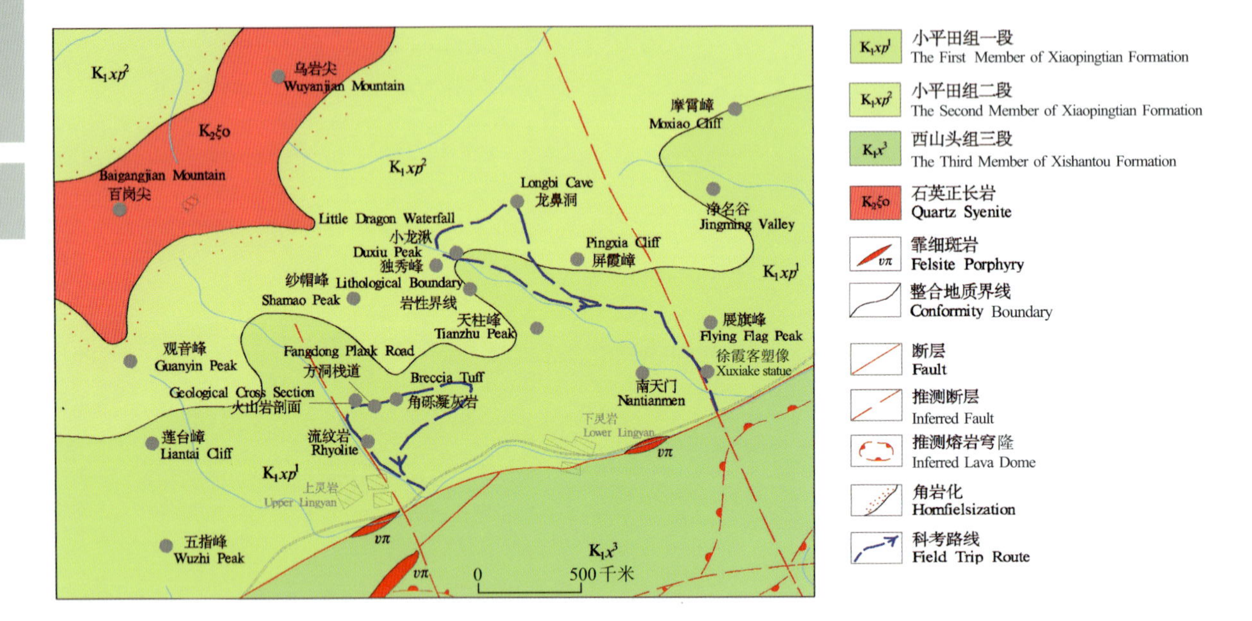

灵岩-方洞景区考察路线图
Field Trip Route of Lingyan-Fangdong Scenic Area

考察强度中等偏高，灵岩景区栈道曲折陡峻

The work intensity is above average. The plank roads in Lingyan Scenic Area are winding and steep.

灵岩景区
Lingyan Scenic Area

目的 Purpose	考察灵岩景区独特的流纹质火山岩、火山岩地貌及人文特色 To study the unique rhyolitic rocks, volcanic landform and the cultural elements in Lingyan Scenic Area
位置 Location	雁荡镇西约4.5千米 About 4.5km west of Yandang Town
路线 Route	灵岩景区栈道 Plank road of Lingyan Scenic Area

灵岩景区在雁荡山诸景之中以岩石铮嵘、环境清幽而著称，同属“二灵一龙”之一，被誉为雁荡山之“明庭”。这里有雄伟的天柱峰，阔大的展旗峰，秀锐的卓笔峰，孤拔的独秀峰，古老的灵岩禅寺，幽异奇幻的天聪洞、龙鼻洞、莲花洞，奇幽的龙湫。同时，千百年来的采药文化演变而来的灵岩飞渡表演，是景区的一大亮点，也是浙江省非物质文化遗产之一。

Among the different scenic attractions within the Yandangshan, Lingyan Scenic Area is particularly famous for its spectacular rocks and tranquil environment. It is being regarded as the "bright courtyard" of Yandangshan. The attractions include the magnificent Tianzhu Peak, Flying Flag Peak, Zhuobi Peak, Duxiu Peak, Lingyan Temple, Tiancong Cave, Longbi Cave, Lianhua Cave and the marvellous Little Dragon Waterfall. In addition, the Fly Across Show of Lingyan is a must-see programme. It is accredited as a provincial intangible cultural heritage which has evolved from the tradition of herbal medicine-picking and mountaineering skills.

徐霞客

徐霞客，明代(1587—1641年)，曾3次考察雁荡山(1613年3月31日，1632年3月20日—4月15日，1632年4月28日—5月8日)。查明大龙湫之水并非来自雁湖。写有雁荡山游记两篇，称赞雁荡山奇峰异洞，并评价为：“叠嶂锐峰、左右环向，奇巧百出，其天下奇观。”雁荡山地貌叠嶂、锐峰名称是由徐霞客提出的。

徐霞客塑像
Xu Xiake Statue

Xu Xiake

Xu Xiake of the Ming Dynasty (1587 - 1641) had visited Yandangshan three times (March 31, 1613; March 20, 1632 - April 15, 1632; April 28, 1632 - May 8,1632). He concluded that the water of Giant Dragon Waterfall is not from Yanhu Lake. There are two travelogues about Yandangshan to praise marvellous peaks and caves of Yandangshan as "There are cliffs(layer upon layer of cliffs)and mountains in different shapes, weird and wonderful in the world". The names of cliffs and mountains of Yandangshan were mostly named by Xu Xiake.

展旗峰

展旗峰与天柱峰相对，高约260米，状如展开的旗帜，故名。此峰为典型的流纹岩层，其上如水流般的纹理，记录了距今约1亿年前岩浆流动的痕迹。

Flying Flag Peak

With a height of about 260m and opposite to Tianzhu Peak, this peak looks like an opened flag. The rocks are rhyolites with very typical flow-banding structure which is a strong evidence of previous lava flow about 100 million years ago.

展旗峰
Flying Flag Peak

屏霞嶂

屏霞嶂高120米，东西宽250米，壁立千霄、五彩相间，状如锦屏。远远观之，壁上流纹构造、球泡角砾构造清晰，记录了1亿多年前火山喷溢、岩浆流动的形态。

Pingxia Cliff

Pingxia cliff is 120m in height,and from east to west is 250m wide. Being blazing with color,the thousand-metered cliff stands just like a screen.Seen from a far distance,the flow-banded structure,lithophysal and breccia on the cliff, which have recorded the volcano eruption and magma flowing morphology about 100 million years ago,are clear.

屏霞嶂
Pingxia Cliff

灵岩禅寺

始建于北宋太平兴国四年(公元979年)，为十八古刹之首堂，寺因灵岩得名。禅寺元末毁于兵火，明清年间，多次毁坏，复而重建，直至民国。近代禅寺再次被毁，于1998年再次重建，可谓一波三折，终幸得存。

Lingyan Temple

Lingyan Temple was built in A.D. 979 (Northern Song Dynasty) and was named after Lingyan Peak. It is the most important temple among the eighteen temples in Yandangshan. The temple was destroyed and rebuilt several times in late Yuan Dynasty as well as in Ming and Qing Dynasties until the establishment of the Republic of China. It was again destroyed and rebuilt in 1998.

灵岩禅寺
Lingyan Temple

天柱峰

天柱峰峰高270米，宽250米。清代喻长霖有楹联曰“左展旗，右天柱，后屏霞，数千仞，神工鬼斧，灵岩胜景叹无双”。其形态由流纹岩受断裂、风化剥落塑造而成，峰壁上显示了岩浆活动留下的纹理及球泡。天柱峰与展旗峰之间有飞渡表演，堪称一绝。

Tianzhu Peak

Tianzhu Peak has a height of 270m, a width of 250m and was described by the famous poet Yu Changlin of Qing Dynasty as a unique natural wonder when compared with other landscapes. The spectacular shape was formed by faulted, weathered and eroded rhyolitic peaks. The flow banding and the globular structures in the upper part of the peaks are traces of previous magmatic activities. The Fly Across Show between Tianzhu Peak and Flying Flag Peak is an amazing traditional herbal medicine picking skill.

天柱峰
Tianzhu Peak

龙鼻洞

龙鼻洞宽10米，深30米，高40余米，洞尽端有石柱连顶，四处透空，似龙鼻，故名龙鼻洞。该洞由于两条断裂作用使岩石开裂、岩块崩落而成。洞顶有岩浆呈脉状（闪长岩脉）侵入于火山喷溢的流纹岩层。闪长岩脉石质细腻，古人科学地选为石刻之处。洞壁80余处的历代摩崖石刻，有“雁山碑窟”之称。沈括、徐霞客等均考察过此洞，为国家级文物保护。

Longbi Cave

The cave is 10m wide, 30m deep and over 40m high. With stalactite and stalagmite connected to the top at the end part of the cave. The cave is well-ventilated. As it looks like a dragon nose and therefore has been called Longbi (dragon nose) Cave. The cave was formed by two faults. There are diorite veins intruded into the effused rhyolitic layers at the top of the cave. The veins are fine and smooth, making them ideal spots for rock carving. With more than 80 cliff stone carvings of the past dynasties, the cave wall has long been known as the "cave of inscriptions" of Yandangshan. Famous scholars such as Shen Kuo and Xu Xiake have visited the cave. It has now listed under the national cultural relics protection.

龙鼻洞
Longbi Cave

小龙湫

小龙湫悬崖环峙，瀑飞崖上，触石腾空，如长龙饮涧。瀑高约70米，不足大龙湫之半，故名。瀑壁为火山喷溢的流纹岩，受断裂影响，同时流水常年冲刷，岩石不断风化崩落，形成了瀑下陡峭的山谷。

Little Dragon Waterfall

This waterfall looks like a flying dragon drinking water in the plunge pool. Its height is 70m, about half of the height of Giant Dragon Waterfall. The rocks are rhyolites from previous volcanic eruption. The water cuts down through a weaker fault line, eroding the nearby rocks and leading to rock fall and development of steep valley.

小龙湫　Little Dragon Waterfall

岩性界线

向对面山峰望去，似乎对面的山峰从中部被水平切割成了上、下两部分。上部为流纹岩，近水平的流纹构造发育；下部为含角砾球泡流纹岩，部分角砾和球泡脱落后在岩石表面形成空洞。二者应为不同火山韵律下形成的岩石，二者之间的分割线为岩性界线，岩性界线近于水平说明该区域构造变形作用微弱。

Lithological Boundary

The peak looks like being horizontally cut to form the upper and lower sections. The upper section is composed of rhyolite with horizontal rhyolitic structure. The lower section is composed of breccia-bearing lithophysal rhyolite. Parts of the breccias and lithophysa are detached due to weathering and erosion to form cavities on the rock surface. The two sections are clearly separated by distinctive lithological boundary indicating their formations during different volcanic periods. The near-horizontal boundary implies the relatively weak regional tectonic deformation of the rocks.

岩性界线　Lithological Boundary

独秀峰

独秀峰峰体方柱形，高100多米，节理裂缝将之与山体分开，使峰体完全独立。峰上近水平的纹理极为清楚，这是火山喷溢岩浆流动的记录。

Duxiu Peak

With a height of over 100m, this square column was separated from the massif by a major joint. The horizontal flow banding is quite obvious. It has recorded the previous magmatic activities in the geological history of Yandangshan.

独秀峰
Duxiu Peak

方洞景区
Fangdong Scenic Area

目的 Purpose	主要考察雁荡山破火山形成的岩石、地层产状和火山岩地貌景观 Mainly to investigate the rock, attitude of stratum and geomorphologic landscape of the volcanic rock that formed by the Yandangshan Caldera
位置 Location	雁荡镇西约6.2千米 About 6.2km west of Yandang Town
路线 Route	上灵岩村至方洞公路及方洞景区栈道 Upper Lingyan Village－Fangdong Highway and Fangdong scenic area plank road

方洞景区位于雁荡山园区的中部，是考察研究雁荡山火山多期次喷发形成不同类型岩石地层单元的典型地点。观音峰自上而下分别反映了4个期次的火山喷发，方洞栈道顺“金腰带”而行，在小的露头尺度上展示了火山作用过程中不同期次喷发形成的火山岩的缩影。沿途奇峰遍布、怪石嶙峋、数不胜数。

It is located near the centre of the Yandangshan Scenic Area. It is an ideal site for field studies and research on the volcanic activities of the Yandangshan Caldera. From top to bottom, the Guanyin Peak displays the four different lithological units of the Yandangshan Caldera. Following the Fangdong Plank Road along the ‘Golden Belt’, one can easily see the outcrops demonstrating the different characteristics of volcanic eruption of the Yandangshan Caldera. Spectacular peaks and rocks of grotesque shapes are widespread and countless.

流纹岩

从上灵岩村步行至方洞停车场，主要考察雁荡山破火山第二期火山喷溢的流纹岩层，即破火山复活期的熔岩流——巨厚流纹岩层，地貌上形成嶂、洞。金带嶂下部和观音峰莲台以下的岩石均为流纹岩。

Rhyolite

Walking from Upper Lingyan Village to Fangdong parking lot, the main purpose is to investigate the rhyolite formation, lava flow and revival of volcano during the second stage volcanic eruption of Yandangshan Caldera. The geomorphologic features comprise the ultra-thick rhyolite formation, cliffs (screen-like peaks) and caves. The rocks at the lower part of the Gold Band Peak and below the lotus throne of Guanyin Peak are all rhyolites.

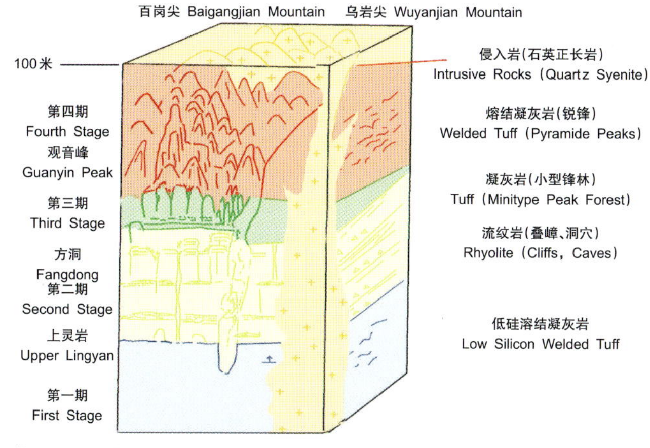

雁荡山各期次岩石与地貌
Rocks and Landforms of Yandangshan in Various Stages of Volcanic Eruption

观音峰

在方洞停车场西望观音峰，观察雁荡山破火山第一、第二、第三、第四期喷发产物的地貌形态。

(1)上灵岩村公路南侧山体为雁荡山火山第一期火山爆发的低硅熔结凝灰岩(K_1x^3)。

(2)观音峰莲座以下至上灵岩村，雁荡山火山第二期火山喷溢的巨厚流纹岩层(K_1xp^1下部)。

(3)观音峰莲座为雁荡山火山第三期火山爆发凝灰岩、熔结凝灰岩并夹有流纹岩层(厚2~10米，向北微倾近水平的岩层)呈莲瓣状(K_1xp^1上部)。

(4)观音"座身"呈锐峰，为雁荡山火山第四期火山爆发的熔结凝灰岩(K_1xp^2)。

Guanyin Peak

The Guanyin Peak can be viewed from west of Fangdong parking lot. It is also an ideal site to observe the four different stages of eruption of Yandangshan Caldera.

(1) The mountain at the south side of the highway of Upper Lingyan village is composed of low silicon welded tuff (K_1x^3) formed during the first volcanic eruption of the Yandangshan Caldera.

(2) From the lotus throne of the Guanyin Peak below to Upper Lingyan Village, the ultra-thick rhyolite formation (K_1xp^1 below) was present. It was formed during the second stage of volcanic eruption of the Yandangshan Caldera.

(3) The lotus throne of the Guanyin Peak is made up of ash tuff, welded tuff and interbedded with rhyolite of 2–10m thick and is slightly tilted towards north with nearly horizontal strata. They were formed during the third stage of volcanic eruption of Yandangshan Caldera. The formation resembles the shape of lotus pedals (K_1xp^1 above).

(4) As for the Guanyin Peak, the lotus throne of Guanyin in sharp peak shape is welded tuff (K_1xp^2) formed during the fourth stage of volcanic eruption of Yandangshan Caldera.

观音峰 Guanyin Peak

火山岩剖面

此处是直观展示雁荡山火山第三期爆发形成的火山岩的最佳处。由下至上依次为火山爆发从空中降落的火山灰堆积形成的凝灰岩；火山岩浆溢流形成的流纹岩夹层，厚3~4米；火山爆发的火山碎屑流，其底部为地面涌流（又称干涌流），具有水平层理。

Volcanic Rock Section

This spot is the best view point to see the volcanic rocks formed during the third stage of the volcanic eruption of the Yandangshan. From the bottom to the top, the formations arranged orderly first with the tuff formed by accumulated volcanic ash then followed by a 3–4m thick rhyolite formed by lava flow. Overlying this rhyolite stratum was welded tuff formed by pyroclastic flow which was underlaid by surge tuff formed by horizontally bedded ground surge (or called dry surge).

火山岩剖面
Volcanic Rock Section

方洞栈道

沿天然岩层开凿栈道，栈道中以“方洞”称胜，故名。栈道的岩壁，即断面，主要为凝灰岩和一层流纹岩，二者接触面在栈道中段最清楚。栈道沿途风景优美，景色各异，随处可见各类典型的火山地质现象。

Fangdong Plank Road

The plank road was built along natural rock formation, in which, "Fangdong" was the most perfect place, so it was named Fangdong Plank Road. The rock wall of the plank road, namely the section, is dominated by tuff and a layer of rhyolites. The contact surface of the two is the most clear at the middle part of the plank road. There are different beautiful sceneries along the plank road, and various types of typical volcanic geologic phenomena can be found everywhere.

方洞栈道
Fangdong Plank Road

角砾凝灰岩

雁荡山经历了多期次火山作用，每期次形成的火山岩存在较大差别。此景点虽小，却显示出3期次火山活动。最早期次的灰白色气孔-杏仁状流纹岩成为稍晚期次深灰色玄武-安山岩的角砾，而它们二者又成为更晚期次的外围灰白色晶屑凝灰岩的角砾，这说明早期固结的岩石被再次的爆发活动炸裂成碎块并再次固结成岩。

Breccia Tuff

Yandangshan has experienced several volcanic eruptions and there are great differences among volcanic rocks formed at each time. This geosite is small but is able to display the three different stages of volcanic activities. The earliest gray-and-white stomata-amygdaloidal-shaped rhyolites become the breccia of later dark gray basalt-andesite rock, and both of them become the breccia of outer off-white crystal tuff of the followed later stage, which indicates that the early consolidated rock was burst into fragments by another eruption activity and was consolidated into rock again.

角砾凝灰岩　Breccia Tuff

大龙湫景区

Dalongqiu Scenic Area

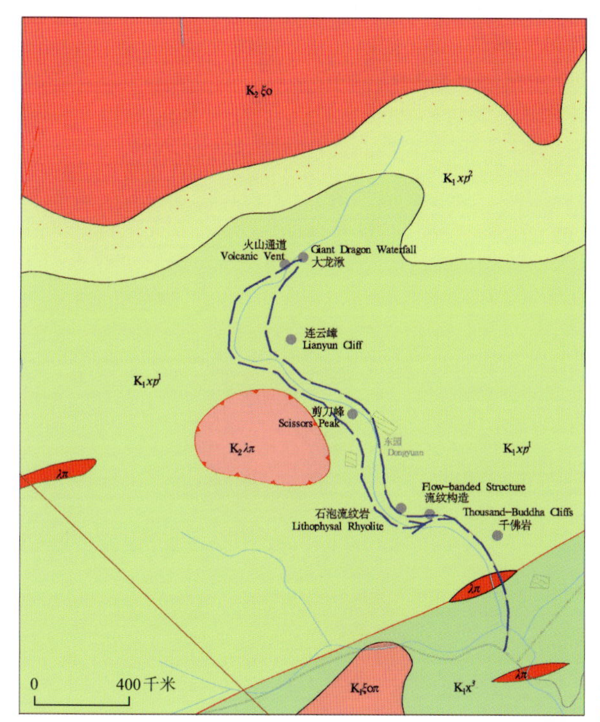

大龙湫景区考察路线图
Route for Field Trip of Giant Dragon Waterfall Scenic Area

大龙湫景区
Giant Dragon Waterfall Scenic Area

目的 Purpose	考察大龙湫景区独特的流纹质火山岩、火山岩地貌及瀑布景观 To investigate the unique rhyolitic volcanic rocks, landform as well as the waterfall landscape in Giant Dragon Waterfall Scenic Area
位置 Location	雁荡镇西约8千米 About 8km west of Yandang Town
路线 Route	大龙湫景区栈道 The plank road of Giant Dragon Waterfall Scenic Area

大龙湫景区在“雁荡三绝”中独占鳌头，在宋代便有“龙湫一派天下无，万众赞扬同一舌”的美誉。沿溪溯徐霞客之古道，有流纹岩构成的剪刀峰、大龙湫、连云嶂诸景，溪东有常云峰、千佛岩，过千佛岩可至龙湫背。一路奇峰绝壁，飞瀑如龙，美轮美奂，令人惊奇。

Giant Dragon Waterfall Scenic Area is one of the three best scenic sites in Yandangshan. It has been highly praised by people since Song Dynasty. Tourists can follow the ancient travel route of Xu Xiake by viewing the important attractions such as Scissors Peak, Giant Dragon Waterfall, and Lianyun Cliff. To the east of the waterfall are Changyun Peak and Thousand Buddha Cliffs which lead to Longqiubei (Back of the Dragon) Mountain. Along the way, the scenery of marvelous peaks, cliffs and waterfalls are beautiful and amazing.

千佛岩

千佛岩山石高低有致、凹凸参差，形态各异，如石佛聚会，故名。雁荡山多期次火山活动喷发溢流的岩浆在冷凝收缩过程中形成的裂隙，受风化作用及流水侵蚀而形成了雁荡山这种独特的叠嶂景观。

Thousand-Buddha Cliffs

These staggered rocks of different sizes and heights are arranged systematically in the area. Joints created by lava cooling during the multiple eruption of Yandangshan allowed weathering and erosion to take place and disintegrated the rock mass into many smaller blocks. Many of these smaller blocks resemble a group of Buddha sitting and attending a meeting. This unique feature characterizes the clustered cliffs and rocks of Yandangshan.

千佛岩　Thousand-Buddha Cliffs

流纹构造

流纹构造是由于岩浆在流动过程中不同颜色、不同成分的物质定向排列所形成的，流动记录了火山喷溢出的岩浆在地表流动的痕迹。

Flow-banded Structure

The rhyolite structure is formed during lava flow by the orientation of lava with different colours and compositions. The flow structure recorded the evidences of volcanic effusion which had brought magma to the land surface.

流纹构造　Flow-banded Structure

剪刀峰

孤峰矗立，峰上部一分为二，状如指向蓝天的剪刀，名曰剪刀峰。此峰是体验移步换景的典型，随着进入，此峰依次出现“剪刀”“啄木鸟”“熊岩”“一帆”和“桅杆”等造型的变化。峰体为流纹岩受断裂影响，经风化作用和流水侵蚀而成孤峰。

Scissors Peak

This solitary peak looks like a pair of scissors standing upright from the ground. It was formed after the collapse of weathered and eroded rocks along faults. When viewed from different positions, the peak may look different and change its shape from scissors to woodpecker, bear, sail or mast.

剪刀峰　Scissors Peak

啄木鸟峰　Woodpecker Peak　　熊岩　Bear Rock　　一帆峰　Sail Peak　　桅杆峰　Mast Peak

球泡流纹岩

雁荡山几乎所有岩石都是由火山喷发而形成的，喷出物都为酸性物质。岩石表面的许多“石球”地质学称“球泡构造”。它是酸性熔岩的表面由于凝固时气体逸出、体积缩小而产生的具有空腔的多层同心圆球体。每一球层常由放射状纤维钾长石或长英质矿物组成。球泡见于玻璃质岩中，尤以在黑曜岩、流纹岩中最为常见，空腔内常被微细的次生石英、玉髓等矿物替代。

Lithophysal Rhyolite

Almost all of the rocks in Yandangshan were formed by volcanic eruption with acidic substances. Many "stone balls" found in the rocks are geologically called "lithophysa structure". It is formed by gas contained and aggregated in certain parts of the magma and lava during flowing. They were later cooled and solidified to form "balls" and these "balls" were empty at the centre due to the presence of gas. Each layer was often composed of radial fibrous potassium feldspar or felsic minerals. Lithophysae were found in vitreous rocks especially in obsidian and rhyolite. The empty cavities were very often replaced by fine secondary quartz, chalcedony and other minerals.

球泡流纹岩　Lithophysal Rhyolite

火山通道

远处的山崖石壁表面可以观察到流纹构造，但它不是水平状的，而是陡立的，表明这里靠近岩浆溢出地表的通道，大龙湫附近的流纹岩便是从这里喷溢而出的，这里看到的是火山通道残留的部分。

火山通道是岩浆从岩浆库穿过地下岩层经火山口或溢出口流出地面的路径，火山喷出的大量物质就是经这些通道运移至火山口而溢出地面的。与主要通道相连的还有许多无固定形状的分支通向地面，或在地下尖灭而消失。火山通道的形状与火山喷发的类型有关。中心式喷发的常具有一个主要的通道，铅直方向，似圆筒状，一般称之为火山筒或火山管。裂隙式喷发型的通道常呈长条状或不规则状。

Volcanic Vent

The flow-banded structure here is in a vertical rather than horizontal fom. This indicates the location of volcanic vent where magma had once erupted though here. This is also the origin of the rhyolites found in Giant Dagon Waterfall. What we can see now are the remains of this important volcanic vent and dome.

Volcanic vent is the route where magma from underground magma chamber extrudes to the surface. Large quantities of volcanic materials are delivered through this passage to the ground surface. There are still many branches in different shapes connected to the main channel. They lead to the ground or become thinner and even disappeare underground. The shape of the volcanic channel is related to the type of volcanic eruption. Central vent eruption usually has a main channel which is in vertical direction and cylinder shape. Generally, it is called volcanic pipe or volcanic chimney. The channel of fissure eruption type is usually in long strip shape or irregular shape.

火山通道　Volcanic Vent

大龙湫

一嶂拔地，水自天来，如蛟龙饮涧，故名。高197米，为国内四大名瀑之一。中国著名地理学家徐霞客(1587—1641年)曾3次来到这里探寻研究，并将这些经历记录了下来，查明大龙湫水的源头为龙湫背，而不是来自雁湖，纠正了史书上有误的记载，被后人收录于《徐霞客游记》中。而后文人墨客络绎不绝，岩壁留下了摩崖石刻20多处。

Giant Dragon Waterfall

The water seems falling down from the sky, whereas the vertical peak looks like a water-drinking dragon. Giant Dragon Waterfall (also Dalongqiu) has an elevation of 197m and is one of the four most famous waterfalls in China. Ancient Chinese geographer Xu Xiake (1587–1641) visited here three times and compiled details of his journeys into his book of *The Travel Book of Xu Xiake*. He made efforts to find out that Giant Dragon Waterfall was actually originated from Longqiubei Mountain rather than from the Yanhu Lake. He also corrected the erroneous records in history books. After Xu, many literati visited here and left more than 20 rock inscriptions on the cliffs.

俯瞰大龙湫 Overlooking Giant Dragon Waterfall

石桅岩景区

Shiwei Peak Scenic Area

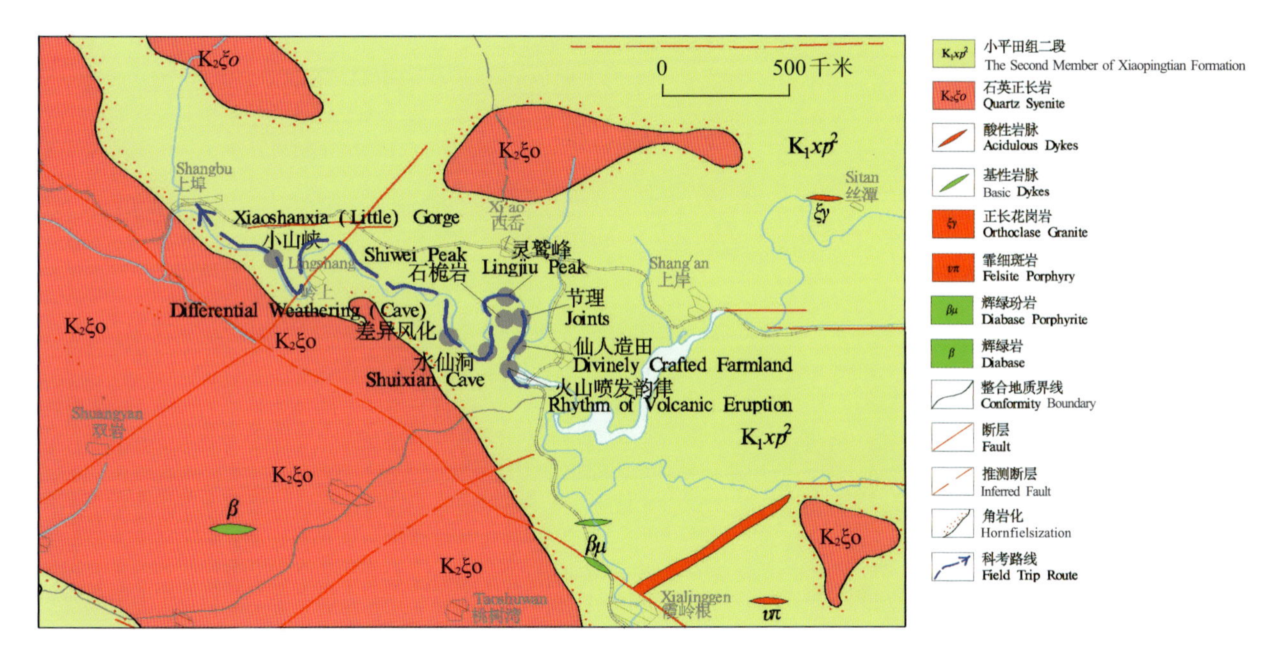

石桅岩景区考察路线图
Route for Field Trip of Shiwei Peak Scenic Area

有一段较陡的山腰栈道

There is a rather steeper plank road along the mountainside.

石桅岩景区
Shiwei Peak Scenic Area

目的 Purpose

主要考察石桅岩及周围火山岩、火山岩地貌，考察地貌与构造作用、风化作用的关系

To study the relationship among Shiwei Peak and the surrounding volcanic rocks, landform, tectonism and weathering

位置 Location

永嘉县东北约33.3千米

About 33.3km northeast of Yongjia County

路线 Route

从永嘉县城经S223或S26到达枫林镇后经珍上线、雁楠公路到达石桅岩景区，或者从雁荡镇经雁楠公路到达石桅岩岩景区，再沿景区内部道路游览

This can be accessed from Yongjia County through S223 or S26 to Fenglin Town. Passing through Zhen Xikou-Shang Butou line and Yannan Highway to Shiwei Peak Scenic Area. Alternatively, it can also be reached from Yandang Town and going through Yannan Highway to Shiwei peak Scenic Area.

石桅岩景区是雁荡山世界地质公园楠溪江园区的主要组成部分，东与雁荡山园区毗邻。雁荡山的阳刚之美与楠溪江的阴柔之美，在这里得到了完美统一，景区集"雄、奇、险、秀、幽"多种特征于一身，是楠溪江园区地质遗迹最为集中的区域，总体反映出在断裂发育和地壳抬升的背景下，流纹质火山岩经长期流水侵蚀所形成的孤峰和峡谷地貌，是典型的流纹质火山岩地质地貌景观。主要景点有石桅岩、小山峡、水仙洞、麒麟峰、将军岩、象岩和岭上古村等。

Shiwei Peak Scenic Area is an iconic attraction of Nanxijiang Scenic District of Yandangshan UNESCO Global Geopark. It is located to the west of Yandangshan Scenic Area. The unique charm of Yandangshan and the delicate beauty of Nanxijiang are perfectly integrated in this area. The area has often been described as "majestic, peculiar, precipitous, graceful, and quiet". It is an area with the most concentrated geosites in Nanxijiang Scenic Area, Shiwei Peak Scenic Area features isolated peaks and gorges formed by rhyolites, aided by crustal uplifts and shaped by river erosion. The main geosites are Shiwei Peak, Xiaoshanxia Gorge, Shuixian Cave, Qilin Peak, Jiangjun Rock, Elephant Rock and Lingshang Ancient Village.

石桅岩

石桅岩高306米，通体红润，孤峰独傲，因远眺似桅帆，故名。它是楠溪江园区的标志性景观。约1亿年前，火山喷出的岩浆冷却后形成火山岩，最初的石桅岩是高耸的火山岩台地，地质构造在石桅岩的形成过程中起到了重要作用，厚度巨大的火山岩被该区3组断裂交会切割形成破碎带，发生崩塌及流水侵蚀，由边缘向中心不断风化缩小，其蚀余残留突出于地表，形成似桅杆的柱状孤峰。

Shiwei Peak

Shiwei Peak is 306m high with a rosy body. It looks like a sail and therefore has been called Shiwei Peak ("shi"means rock and "wei" means sail in Chinese). It is an iconic geosite of Nanxijiang Scenic Area. About 100 million years ago, lava from volcanoes flowed and cooled to form rocks. The original Shiwei Peak is a towering lava plateau. Geological structure has played a significant role in the forming process of Shiwei Peak. The volcanic rock of huge thickness was cut by three fault segments of this zone to form fracture zone, which was weathered and reduced continuously from edge to centre in case of suffering collapse and water erosion, with residuals protruded to the earth's surface to form mast like columnar isolated peak.

石桅岩
Shiwei Peak

火山喷发韵律

火山喷发韵律是火山喷发出的碎屑物按颗粒从大到小、密度从大到小的顺序先后分层沉积而成岩层的规律。靠近竹林深处的崖壁，熔结凝灰岩如夹心饼干一般一层层叠加，它是火山喷发韵律的良好露头。由于火山在多次喷发过程中，每次喷发的物质成分和规模有所不同，导致火山岩有不同的颜色和厚度。

Rhythm of Volcanic Eruption

The rhythm of volcanic eruption is the regular pattern of rock formation which is formed through stratified sedimentation in the order from big to small of particle and proportion. On the cliff wall close to the deep bamboo forest, welded tuffs are overlapped layer by layer just like sandwich biscuits, which is a clear outcrop demonstrating the rhythm of volcanic eruption. Since the material compositions and scales of each eruption varied in the process of multi-times volcanic eruptions, the volcanic rocks are different in colours and thicknesses.

火山喷发韵律
Rhythm of Volcanic Eruption

仙人造田

岩石表面纹路纵横交错，形似田埂，故民间称之为“仙人造田”。“田”为熔结玻屑凝灰岩，具凝灰结构和假流纹构造，以玻屑为主，岩屑次之，晶屑很少。玻屑具定向性且已脱玻化。“埂”的岩性基本相同，但其中以岩屑角砾为主，定名为“含角砾熔结凝灰岩”。“田”和“埂”两者岩性虽然没有很截然的区别，但仍然有差异，表明不是同时形成的。

其成因推断为在早期火山碎屑尚未完全固结之前，形成原生的柱状节理。后期裂缝受火山热液下渗及雨水的影响有微弱硅化，从而增强了岩石的抗风化能力，当岩石暴露在地表经受风化剥蚀后，抗风化能力的差异呈现出凹下的“田”和凸出“埂”的形态。

Divinely Crafted Farmland

The lines on the rock surface are criss-crossed looking like footpath of fields, so the folks named it "Divinely Crafted Farmland". The "field" is welded vitric tuff, with tuffaceous texture and pseudo-rhyolitic structure. Among which, vitric fragment is dominated and is followed by debris, and there are very few crystal fragments. Vitric fragment has directionality and has been devitrified. The lithology of the "footpath" is basically the same, but rock debris and breccia are dominated, which is named as "breccias-bearing welded tuff". The lithology of the "field" and "ridge" is not very different but the difference still exists, indicating that they are not formed at the same time.

It is inferred that the primary columnar joints formed before the early volcanic clastic was not fully consolidated. Affected by the volcanic hydrothermal infiltration and rainwater, the later cracks are slightly silicified thereby enhancing the weathering resistance of the rock. After the rock is exposed to the surface and subjected to weathering and erosion, the difference in weathering resistance will be presented in a way in which the concave is "field" and the bulge is "footpath".

仙人造田
Divinely Crafted Farmland

节理

这里坚硬的岩石受到地质作用而产生裂隙，裂隙间岩石并没有产生明显的错动，地质学上把这些裂隙叫作节理。如果岩石中发育大量节理会使岩石疏松易碎，引发崩塌。节理密集发育的地方是建桥修路过程中的重点关注对象。节理按成因分为原生节理和次生节理两大类，原生节理是指成岩过程中形成的节理，例如沉积岩中的泥裂，火山熔岩冷凝收缩形成的柱状节理，岩浆入侵过程中由于流动作用及冷凝收缩产生的各种节理等。次生节理是指岩石成岩后形成的节理，包括非构造节理（风化节理）和构造节理。

Joints

The hard rocks fractured here without obvious physical displacement of their structure. The fractures are called joints. When joints are abundant, the rocks can easily break and will cause rockfall and landslide. Much attention should be paid to sites with high concentration of joints as they are of great concern to bridge construction and road maintenance. According to the cause of formation, joints can be classified into two categories, including primary joint and secondary joint. Primary joint refers to the joints formed in the diagenesis process, such as the mudcrack in sedimentary rocks, the columnar jointing formed by condensation and contraction of volcanic lava, and various types of other joints generated due to flow, condensation and contraction in the magmatic intrusion process. Secondary joint refers to the joints formed after rock formation, including non-tectonic joint (weathering joint) and tectonic joint.

节理　Joints

灵鹫峰

两座并列的山峰，像一只振翅欲飞的灵鹫，故名。这是雁荡山大规模火山活动形成的熔结凝灰岩在水仙洞北西向断层控制下形成陡壁，并在节理裂隙、风化及流水侵蚀下形成两个孤立的锐峰。

Lingjiu Peaks

The two peaks stand side by side like a "holy eagle". They were formed by continual weathering and erosion along faults and joints on the volcanic rocks formed by the extensive volcanic activities in Yandangshan.

灵鹫峰 Lingjiu Peaks

水仙洞

水仙洞是个天然的断层石洞，洞深10余米，高达8米，宽3～5米。洞穴受走向为120°～300°，倾角近于直立的水仙洞断层控制，与控制石桅岩的斜切主断层为同一组断裂。断层形成一定宽度的破碎带后，流水侵蚀及风化作用使破碎的岩石崩落而形成如今的形态。在水仙洞内可见断层挤压形成的擦痕，以及断层带内挤压脆性变形形成的破碎岩块，断层存在分叉的现象，两个滑移面产状分别为220°∠84°和305°∠50°。

Shuixian Cave

Shuixian Cave is a natural fault cave with a depth of more than 10m, with height up to 8m as well as width of 3-5m. Under the control of the 120°-300 ° fault with a nearly upright dip angle, the cave is in the same group of faults together with beveled main fault which controls the Shiwei Peak. After the fault had formed a certain width of fracture zones, broken rock fell due to water erosion and weathering, thus the present shape formed. In Shuixian Cave, striation formed by the crushing of the fault and the fractured rock formed by rock of brittle deformation due to crushing in the fault zone can be found. The fault has bifurcation phenomena and the two slip surface occurrence is respectively 220 °∠84° and 305 ° ∠50 °.

断层滑移面 Fault Slip Surface

水仙洞 Shuixian Cave

差异风化

岩石由于物质成分、结构等不同，抗风化能力存在差异。此处岩腔上部为熔结凝灰岩，抗风化能力略强而凸出；下部为凝灰岩、含角砾凝灰岩，抗风化能力略弱而剥落凹陷。上下岩性不同而且抗风化能力不同体现出了火山喷发韵律。

Differential Weathering

Different composition and structure of rocks have different abilities of resistance against weathering. The cavity was a result of differential weathering with the higher resistant welded tuff forming the upper part and the relatively weaker tuff, breccia-bearing tuff forming the lower section. Different lithology and different anti-weathering ability reflect rhythm of volcanic eruption.

差异风化 Differential Weathering

小山峡

在石桅岩至岭上约3千米长的河段，受一系列北西向近直立断层的控制，整体呈北西走向，流水沿断裂侵蚀，形成多处"V"形谷地貌。"小三峡"为上游构造峡谷的代表性景点，即河谷沿着断层等构造软弱带侵蚀而形成的峡谷，代表了典型的楠溪江上游河流侵蚀地貌。河流下蚀作用、侧蚀作用以及冲蚀作用，形成了岩槛、跌水、壶穴、侧蚀槽、冲刷痕等丰富多样的地质遗迹，是研究流水地貌与构造作用不可多得的野外场所。

Xiaoshanxia (Little) Gorge

The reach from the Shiwei Peak to Lingshang Village with a length of about 3km, is controlled by a series of north-west direction nearly upright faults extending from north to west. The V-shaped valleys were formed by water cutting through the faults. Xiaoshanxia (Little) Gorge is a typical tourist attraction of gorge in the upper reach of the stream. Water cut through the weaker areas to create erosional landform of Nanxijiang. Rich and varied georelics is formed, such as threshold (a more resistant rock layer), plunge pools, pothole, groove and scour mark. Due to vertical erosion, lateral erosion and erosion of the river, it is a nice fieldwork place to study fluvial landform and tectonism.

小山峡 Xiaoshanxia (Little) Gorge

方山景区

Fangshan Scenic Area

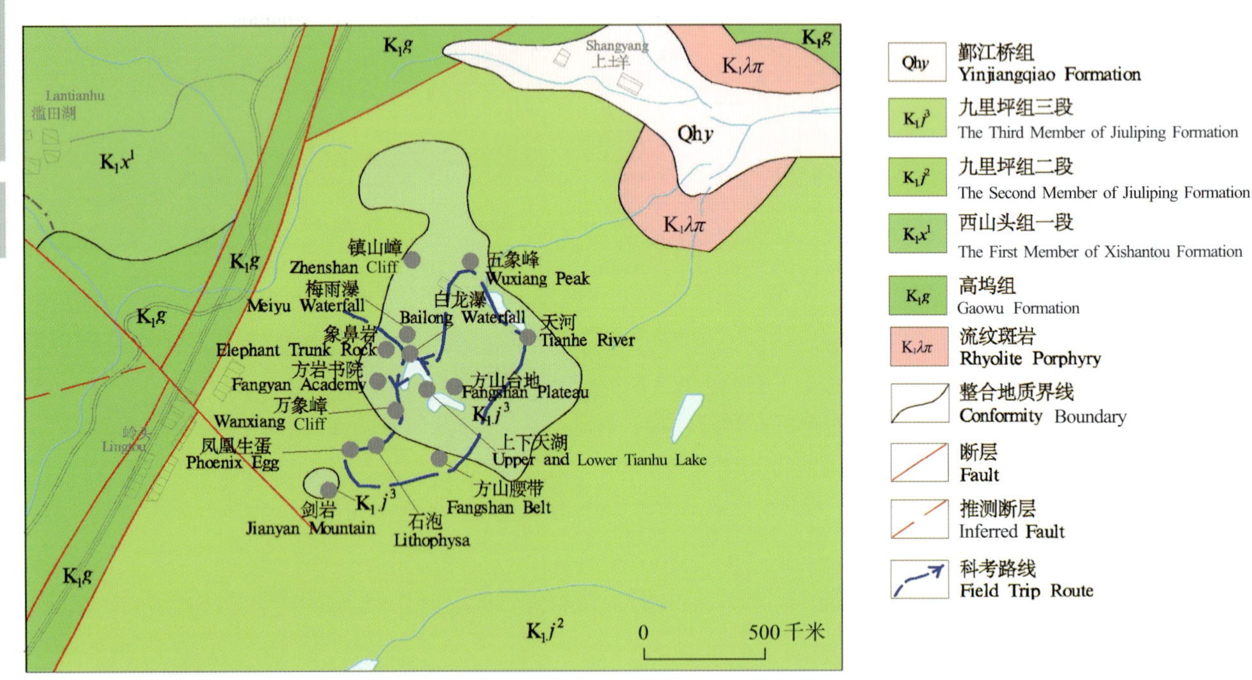

方山景区考察路线图
Route for Field Trip of Fangshan Scenic Area

需要沿栈道步行至方山顶
It is necessary to walk to the top of the Fangshan along the plank road.

方山景区
Fangshan Scenic Area

目的 Purpose

主要考察方山火山岩岩相及岩性特征、岩相岩性特征与地貌的关系、构造与地貌的关系以及其他相关的火山地质遗迹

Mainly to investigate the volcanics lithofacies and lithologic characteristics of Fangshan and their relationship withlandform, tectonic movement and other geological heritages

位置 Location

雁荡镇东北约12.4千米

About 12.4km northeast of Yandang Town

路线 Route

从雁荡镇经G104可到达景区,再沿景区内部道路游览

The Scenic Area can be reached from Yandang Town via G104, and then you can tour along the internal road of scenic area.

方山为典型的复活破火山口构造,山体由流纹质火山岩平台地貌和沟谷溪瀑两部分组成。方山巍峨雄伟,四周被50~100米高的绝壁围限,形成一个千亩高山平台,从下仰望,屏开嶂合,奇峰罗列,瀑流飞泻,气象万千。置身其上,温黄平原风光尽收眼底。红日初升,波光跃金,天海相映,壮丽之极。更有云海佛光、日月同升和瀑布倒流,堪称“方山三绝”。

The geology of Fangshan is a typical revived caldera. The mountain is made up of rhyolite volcanic rock plateau landform, valley, creeks & falls. Fangshan is surrounded by cliffs with height of 50-100m which are parts of the thousand acres of alpine plateau. When looked up from the foothill, the magnificent screen peaks, undulating peaks and cascading waterfalls can all be seen. The panoramic view of Wen-Huang Plain can be viewed from a higher ground which is particularly attractive during sunrise with red, glistening golden light and the sea and sky setting. The sea of clouds, Buddha's light, simultaneous sunrise with moon as well as the reversed fall are combined to be called "The Three Incomparable Sceneries of Fangshan".

镇山嶂

镇山嶂是方山西侧悬崖的总体景观,悬崖高50多米,蜿蜒曲折地展开,凹处形成幽谷。远观可见雁荡群山磅礴之气势,也可入幽谷近仰方山绝壁之巍峨。岩壁上有“镇山”两个大字而得名镇山嶂。方山四面悬崖形成的各种嶂状景观是熔岩台地地貌的基本表现。

Zhenshan Cliff

It is located in the western side of the Fangshan, which has a height of over 50m, from where an overview of the mountains of Yandangshan can be seen. When being viewed from a distance, the majestic setting of Yandangshan are marked by valleys and lofty cliffs. There are two Chinese characters "镇山"(Zhenshan) carved on the palisades. The various screen-shaped landscapes formed by cliffs on the four sides of the Fangshan are the basic representation of the lava plateau landform.

镇山嶂
Zhenshan Cliff

五象峰

五象峰位于方山北侧悬崖。100米宽的悬崖被垂向节理切割，产生群象并立的景观，又称五象岩、五象嶂。

Five Elephants Peak

It is located in the northern part of Fangshan. The peak about 100m wide is cut by vertical joints to form a group of elephants and therefore has been called Five Elephants Peak.

五象峰
Five Elephants Peak

白龙瀑

白龙瀑位于方山悬崖西侧。白龙峡冲沟下端的瀑布为方山八瀑中流量最大者。瀑布自80米高的峭壁下坠，颇为壮观。瀑下白龙潭，石为底，水浅见底，泄出成小溪。潭广约50平方米。

Bailong Waterfall

This waterfall is located on the cliff on the western side of Fangshan. The waterfall at the bottom of the Bailongxia Gully is the largest among the eight waterfalls in Fangshan Scenic Area. It has a height of 80m with about 50m^2 Bailong Pool underneath where water is shallow and clear enough to see its bottom.

白龙瀑
Bailong Waterfall

天河
Tianhe River

天河

天河是一条沿北西向节理发育的峡谷，长300米，宽30米，深25米，两侧悬崖直立，筑坝蓄水成为悬河奇观。又因河谷如刀切般平直，似王母玉簪划开，因而引发一些美丽的民间传说。天河是沿节理带风化崩塌作用的最终结果。进一步发育将促使平台的分离和孤峰的形成。

Tianhe River

Tianhe River is a gorge developing along a north-west trending joint. It is 300m long, 30m wide and 25m deep. Cliffs at the two sides of the valley with the dammed river have formed a scenic area. It is straight like a knife. The straight line is said to be drawn by Mother of Heaven with her jade hairpin (romantic love story in Chinese folktales). Weathering and denudation with the joint are the major formation factors. As a result, the plateau has been continually disintegrated and separated, eventually forming isolated peaks in the area.

方山台地

方山表现为一个面积近0.7平方千米、厚约50米的岩石台地，平缓地分布于相对高差约400米高的山顶之上。构成方山平台的是九里坪组上部的沸溢相碎斑流纹岩，碎斑流纹岩的坚硬致密、不易风化的特征，是后期地质作用过程中发育多组破裂面，崩塌形成方山台地地貌的重要基础。

Fangshan Plateau

The rocky plateau of Fangshan is 50m thick covering an area of nearly 0.7km^2 at a relative height of 400m. The Fangshan Plateau is composed of porphyritic rhyolite belonging to the upper part of Jiuliping Formation. The characteristics of porphyritic rhyolite are hard and resistant to weathering and erosion. The present appearance of Fangshan is a result of a long period of continual denudation along faults and joints.

方山台地　Fangshan Plateau

上下天湖

上下天湖位于方山顶南西侧。在方山平顶波缓开阔的凹地上筑坝蓄水形成两个小水库，称为上下天湖。两个天湖镶嵌在平滑光洁的裸露岩丘之间，清爽明媚，为方山增添了灵气。

Upper and Lower Tianhu Lake

It is located in the southwestern part of Fangshan. The two small reservoirs are formed by damming and impoundment on the flat, wide and sunken place in the flat top of Fangshan. They are called the Upper and Lower Tianhu Lakes. The two lakes are enclosed by surrounding rocks creating an enchanting site for visitors.

上下天湖　Upper and Lower Tianhu Lake

石泡

岩性为碱性流纹质凝灰熔岩，在风化面上，可见到大量石泡，部分石泡破裂成为空壳。这是火山喷溢、岩浆中含有大量气体而没有排逸出来形成的。

Lithophysa

The lithology is alkaline rhyolitic tuff lava. A large number of lithophysas can be found in the weathered surface, and partial lithophysas split to form vacant shell. It is formed by lots of gas contained in lava retained in the formation due to volcanic eruption.

石泡　Lithophysa

凤凰生蛋

沿近东西向劈理发育的刃状尖峰群，夹持着一已风化成椭球体的数米长的巨石，巨石似蛋，出露面积约50平方米。

Phoenix Egg

Sharp peaks were developed along the east-west trending faults with the presence of several metres big oval shape boulders covering an area of 50m^2. They are the result of spheroidal weathering of the original bedrocks.

凤凰生蛋　Phoenix Egg

剑岩

剑岩为一流纹质凝灰熔岩构成的50米高之孤立岩峰，四面绝壁，是方山平台崩裂缩小的最终典型地貌。

Jianyan Mountain

The 50m high solitude pinnacle was formed by rhyolitic tuff. It is the remains of the gradual disappearance of the previously larger plateau in the Fangshan area.

剑岩　Jianyan Mountain

方山腰带

从四面悬崖可见50米高的绝壁上发育2~3条厚数十厘米的水平"腰带","腰带"之间是2~3层熔结凝灰岩,每一层都是一次火山喷溢产生的流动单元,流动过程中产生的蚯蚓状流纹和岩流底部的角砾都清晰可见。而"腰带"则是空气中的火山灰在熔岩溢流间隙期降落堆积形成。这一现象的最佳观察点位于方山西侧羊角洞一带。而沿火山岩层面发育的巨大台地地貌居于四面绝壁之上,尤其令人惊奇。

Fangshan Belt

Two to three horizontal belts can be seen 50m up in the cliffs each with thickness of less than a metre. There are two to three layers of welded tuffs. Each layer represents a flow unit generated by a volcanic eruption. The lines of flow banding are generated during pyroclastic flow with breccia and debris settled at the bottom part of the rock stratum. The belt was formed by ash fall during the intermittent period between two lava flows. The best observation point of this phenomenon is the Horn Cave on the western side of Fangshan. The huge lava platform with cliffs all around is particularly stunning.

凝灰熔岩中的大量熔岩条带构成流纹
Flow Banding in Rhyolitic Tuff

方山腰带 Fangshan Belt

长屿硐天采石遗址

Quarry Site in Changyu Dongtian

长屿硐天采石遗址考察路线图
Route for Field Trip of Quarry Site in Changyu Dongtian

采石窟内道路蜿蜒曲折且起伏大
The road inside the caves are winding and fluctuating.

长屿硐天是一座有着1500多年历史的采石矿山遗址，它完整地保留了隋唐至今各个时期留下的采坑岩硐，已知28个硐群，300多个露天采坑与1000多个井下采硐密集分布在1.57平方千米的低丘山地上，采空区体积累计达600万立方米。硐体错落相连，气势恢宏，令人叹为观止。大量的采石遗址景观，完整地再现了古代采石工艺，并在其周边形成了一个广泛运用长屿石材的村落群，具有重要的史学意义。

Changyu Dongtian Cave is a quarry site with over 1500 years of history. It has preserved quarry pits and stone caves of all periods since Sui and Tang Dynasties. It has 28 caverns, over 300 open gravel pits and 1000 underground caves. They are densely distributed in the hills which cover a total area of 1.57km^2. The total volume of earth being removed so far is estimated to be 6 million cubic metres. The caves are interconnected in an amazing complex system. The caves have impressive display of ancient quarrying techniques and their great impacts on the development of quarrying communities in its surrounding area.

石园 Stone Park

目的 Purpose	考察长屿硐天地区石头工艺文化及生活石器 To study the stone culture and tools in Changyu Dongtian region
位置 Location	温岭市东北约10千米 About 10km northeast of Wenling City
路线 Route	石园内步行道 Footpaths in Stone Park

这里展示的200多扇石窗全部都来自当地，是从各居民家里收集而来。石窗是以前建造石屋的时候嵌入在石壁上，用以采光、通风等用，与石屋融为一体。石窗形态各异，寄托了当地的居民对于生活以及未来美好的愿望。石匠将其石板镂空雕刻，造型不同，可分为平安石窗、如意石窗、直棱格石窗、长寿石窗、多子石窗、花篮石窗、梅雀双龙石窗、对花石窗、金钱石窗、蔓草石窗等，每一种又可细分各种不同的造型，体现了当地石匠精巧的工艺和对生活的寄托。

除石窗外，该区域还展示了由长屿石材制作的生产生活设施、健身设施等。

福栖宝瓶
Good Ortune Vase Stone Window

盘龙竖戟
Dragon and Halberd Stone Window

榴叶正茂
Lush Pomegranate Leaf Stone Window

The Stone Park hosts the collection of more than 200 stone windows of local areas. The stone windows were installed in the stone wall when the stone house was built. They were good for letting in daylight and improving ventilation. They come in different shapes which represent the spiritual demands for better life and future of the local people. To manufacture these windows, stone slabs were carved and hollowed to form different patterns and shapes. According to the needs of the people, they can be made and classified into different types such as household safety, good wishes, straight lattice, longevity, many sons, flowers, plum blossom, sparrow, dragons, phoenix and creeping weed. Each type can be further subdivided into a variety of finer patterns and details, reflecting the superb craftsmanship of local stonemasons.

In addition to stone windows, the park also exhibits stone made household appliances and tools, fitness facilities.

用石材修筑的建筑
House Made of Stone

石磨
Stone Mill

水云硐
Shuiyun Cave

目的 Purpose	考察水云硐采石遗址，了解开采方式、过程以及历史 To study the quarry site of Shuiyun Cave and understand the quarry operation and history
位置 Location	温岭市东北约10千米 About 10km northeast of Wenling City
路线 Route	硐外及硐内游步道和栈道 Walking trail and plank road outside and inside the cave

水云硐由52个硐体组成，总面积约1.5万平方米，大部分为现代开采的大型覆钟式采石遗址。其中中国石文化博物馆为我国最大的硐穴博物馆。《神雕侠侣》《鹿鼎记》等多部电视剧剧组曾取景于此，为水云硐增添了亮丽色彩。

Shuiyun Cave (Water and Cloud Cave) consists of 52 smaller caves covering an area of 15,000m^2. Most of them are large inverted bell-shaped caves excavated in modern times. Its Stone Culture Museum is the largest cave museum in China. It was the movie shooting site of many TV series, such as the famous *Legend of Condor Heroes and The Duke of Mount Deer.*

隋唐采石遗址

这是长屿硐天最简单的采石遗址类型：简单的阶坎形，规模较小且孤立分布在裸岩表面，规模小于3米。它是长屿硐天最古老的遗址类型，通过考证，其开凿时代在1500年前的唐代。

Quarry Site of Sui and Tang Dynasties

These ruins represented the simplest method of quarrying in Changyu Dongtian. They were in form of small scale stair and step cutting on bare rock surface of less than 3m wide. They were the oldest quarrying method in Changyu Dongtian dated back to Tang Dynasty (1500 years ago).

隋唐采石遗址
Quarry Site of Sui and Tang Dynasties

崩塌遗迹

大家看这堆巨石，这是现代采石方式造成的崩塌遗址。1997年，山体发生崩塌，崩塌面积约31 380平方米，山体沿西南断层面往下陷落，岩块大者直径百米。

Collapse site

Looked at this pile of boulders, this was the collapse of the quarry site once operated by modern quarrying method. Rock fall and landslide in 1997 had covered an area of 31, 380m². The rocks slipped along the southwestern trending fault plane, releasing large boulders of over 100 metres in diameter to the lower ground.

崩塌遗迹　Collapse Site

石头玩乐室

室内放置了塌落拱、地质罗盘、粗糙度测试仪、迷你石拱桥搭建、迷你石窗雕刻以及各种岩石标本等，墙壁上是园区地质地貌科普宣传画，让游客在游玩中了解地质知识，增加对地球的认知和热爱。

Playing with Stone

Displays such as the collapse arch, geological compass, roughness tester, mini stone arch bridge construction kit, mini stone window carvings, rock specimens, and geological and topographical posters have teaching and learning meanings. They are used by teachers or guides to demonstrate and help to arouse interest of students and visitors in geological and stone features found in the geopark.

石头玩乐室　Playing with Stone

石文化博物馆

石文化博物馆为我国最大的硐穴博物馆，整个博物馆分成采石遗址历史发展区、采石遗址展览区、采石工艺展览区、岩石应用展览区、区域背景区五大部分。这五大部分全面地展示了长屿硐天所有景观的形态特征、形成原因、演化发展，以及它们的突出价值，是园区最重要的采石文化遗址现场博物馆。

采石遗址历史发展区原来是一条长长的运送石料的平巷，这里陈列着长屿硐天从老到新的5种遗址类型，并且把它们与当地社会历史发展的脉络并列展示，以体现遗址类型的发展演化与当时社会历史发展的密切关系；采石遗址展览区包括长屿硐天的遗址景观和浙江省典型的采石遗址景观，其中长屿硐天采石遗址景观由博物馆内部的展板和博物馆本身的实体遗址展示；采石工艺展览区展示传承千年的采石工艺、工具及劳动场景；岩石应用展览区展示古人对长屿岩石的应用，如水利工程、桥梁、民居、生活器物等；区域背景区展示浙江重要的采石区的遗址模型、岩石种类。

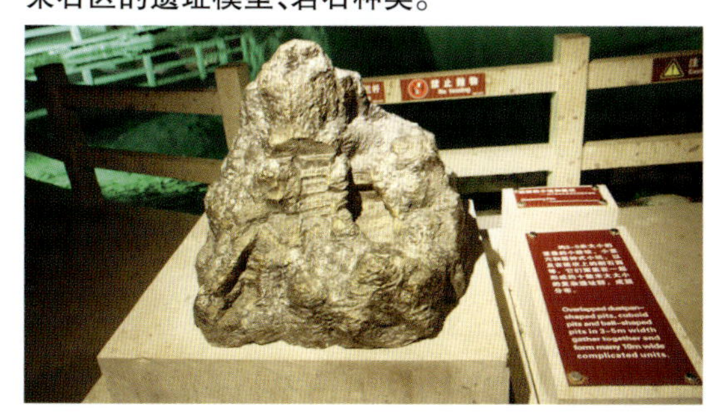

石文化博物馆　Stone Culture Museum

Stone Culture Museum

Stone Culture Museum is the largest cave museum in China. The whole museum is divided into five Sections: Quarrying History, Quarry Site Exhibition, Quarrying Skills and Operation, Stone Application, and Regional Stone Culture. The museum shows the impressive history, operation and development of Changyu Dongtian and has great cultural and historical values.

Quarrying History Section was originally used as a channel for transporting stone slabs. Five relics of Changyu Dongtian are exhibited here, ranging from the oldest to the newest to link quarrying activities with local social and cultural development. Quarry Site Exhibition Section has interpretive boards to display typical quarry site of Changyu Dongtian and other sites found in Zhejiang Province. Quarrying Skill and Operation Section exhibits the operation, tools and the works of labour in the old days. Stone Application Section displays the different applications of the stone excavated in house building as household tools appliances and building materials for bridges, water gates and other infrastructures. A model in the Regional Stone Culture also shows all the important quarries and different types of stones produced in Zhejiang Province.

采石遗址历史发展区　Quarrying History Section

采石遗址展览区　Quarry Site Exhibition Section

采石工艺展览区　Quarrying Skill and Operation Section

岩石应用展览区　Stone Application Section

区域背景区　Regional Stone Culture Section

双门石窟

双门石窟是一处以石窟文化和道教文化为主要特色的旅游胜地。内有十分丰富的摩崖石刻，千姿百态的石窟凿像，精妙深玄的道家故事壁画，是人们访仙问道、避暑休闲的理想去处。双门石窟从明代开始就被宗教使用，石窟岩壁的上部为规模小于3米的阶坎式、直穴式采石遗址，下部为规模小于10米的中型覆钟式采石遗址，石窟内部为规模大于10米的大型覆钟式采石遗址，是一组多时代的复合采石遗址。

Shuangmen Grotto

Shuangmen Grotto is a tourist attraction featuring grotto and Taoist culture. It has very rich wall carvings and finely carved statues to illustrate Taoist story. It is an ideal place for people to visit the immortal, and to study Taoism. It is nice and cool inside, particularly in hot summer days. The grotto was for religious uses since Ming Dynasty. The upper part of the cave is a bench-like perpendicular pit type quarry site of less than 3 meters tall. The lower part is a medium-sized inverted bell-shaped quarry site of about 10 meters tall. Inside the grotto, another large inverted bell-shaped quarry site of over 10 meters can be found. Therefore, this is a set of complex quarry sites which had gone through a long history.

双门石窟　Shuangmen Grotto

观夕硐群
Guanxi Caves

目的 Purpose	考察观夕硐采石遗址，欣赏岩硐音乐盛会和石雕 To study and appreciate excavation and stone carving in quarry sites in Guanxi Cave Group
位置 Location	温岭市东北约10千米 About 10km northeast of Wenling City
路线 Route	观夕硐栈道 Plank road in Guanxi Cave

硐群总体呈北西西走向，由一主平硐自西向东分叉开拓延伸，采坑以多台阶向下采石，由308个硐体组成，面积达5.38万平方米，硐体高低错落，彼此连通，最高处离地百余米，完整地保存了隋唐至今各时期的采石遗址，是采石遗址宏观形态的最佳展示区。有九曲桥、岩硐音乐厅、硐天宝碗、观音石刻等景点。

The caves overall extend from north-west to west, with the main cave extending from west to east. Steps were built to let workers go into the quarry pits. The whole system is composed of 308 caves covering an area of 53, 800m^2. They have different sizes but are interconnected with each other. The highest cave is about 100m. These caves are relics of Sui and Tang Dynasties. The main attractions include Bridge of Nine Turnings, Cave Concert Hall, Treasure Bowl, and Guanyin Sculpture.

硐天花园 Dongtian Garden

岩硐音乐厅 Cave Concert Hall

楠溪江古村落考察路线

Field Trip Route for Ancient Villages in Nanxijiang

楠溪江古村落考察路线图 Field Trip Route for Ancient Villages of Nanxijiang

楠溪江中游自港头至大小楠溪交汇处，沿途不仅可以饱览楠溪江秀丽景色，而且能探寻赏析古村落文化，可以将河流地貌的地质考察与古村落赏析有机结合起来。这些古村落的共同特点：①村落的选址讲究人居环境与山川地势的融合，均以显著山峰作为本村的地标；②建筑布局在美好寓意下，进行一体化设计；③充分利用了天然溪流，实现背山面水的风水佳境实乃生活的便利；④居民具有崇尚耕读文化的传统，村落历史延续久长，名人辈出。

可以说，楠溪江中游的古村落是中国传统乡土建筑的瑰宝，是山水景观与建筑思想的高度结合，耕读文化与宗教文化的充分交融，人类生活与自然环境的无限契合。蕴含在古村布局建筑中的人文精神，可为当今新农村建设提供理念借鉴。楠溪江中游的古村落具有较高的历史价值、科学价值和艺术价值。对于研究我国的人口迁徙史、宗族史、乡村建设史、山水文化都有重要的参考价值。

Ancient villages are located along Nanxijiang from the headwater to the intersection of Dananxijiang and Xiaonanxi River where their general characteristics are shown: ①The villages pay attention to the integration of human settlements and natural landscape, and all the villages take the conspicuous mountain as landmarks of the village. ② Architectural layout is designed based on blessings and good meaning. ③The natural stream is fully used to bring better life and establish beautiful scenery. ④The villagers follow the tradition of the farming-studying culture, these villages therefore have long histories and famous residents.

Nanxijiang Ancient Village is a treasure of ancient rural architecture in China, representing the harmonious integration of architecture with landscape, farming-studying culture and religion as well as humans and natural environment. The spirits embedded in the setting of ancient village inspire modern rural and village design and development. The ancient villages in the middle reaches of Nanxijiang have great historical, scientific and artistic values. They are extremely important as references to the study of the history of migration, religion, village development and landscape culture.

苍坡古村
Cangpo Ancient Village

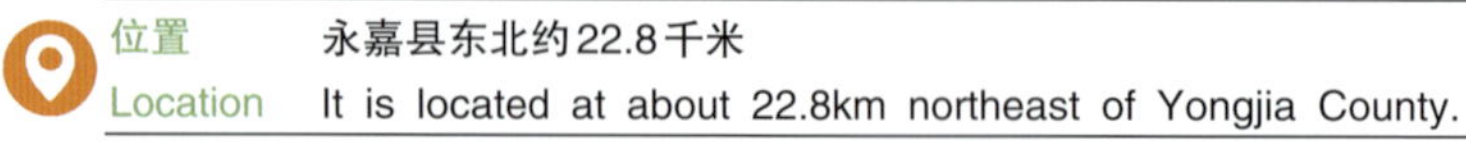

目的 Purpose：主要考察苍坡古村的历史、建筑风格及艺术、与村落有关的人文历史
The field trip is mainly for the study of the history, architecture, art and associated culture of Cangpo Ancient Village.

位置 Location：永嘉县东北约22.8千米
It is located at about 22.8km northeast of Yongjia County.

路线 Route：从永嘉县城经S223到达目的地，再沿村落内部道路游览
The village can be reached through S223 from Yongjia County. Tourism activities can be carried out along the village roads.

苍坡古村，原名苍墩，始建于五代后周显德二年（公元955年），为李姓聚居之地。现存的苍坡村

Cangpo Ancient Village was built in the Later Zhou Dynasty during the Period of Five Dynasties（A.D.955）by a family surnamed Li. The ex-

是九世祖李嵩于南宋淳熙五年(公元1178年)邀请国师李时日按五行风水说，依“文房四宝”布局重建的村落，至今已840多年的历史，平面上略呈方形。

苍坡古村以笔架山为地标，以“文房四宝”而布局展开。直指笔架山的东西向笔直的中心街称为“笔街”，笔尖直指西面“笔架山”(又称“文笔峰”)，建东、西两池为“砚池”，笔架山的倒影在砚池中，意为“文笔蘸墨”。砚池边沿用条石砌筑砚槽，双池边靠近笔街的地方各放置一大条石，意为墨锭，方正的村落地块以及农田如同一张田字格的白纸，期待后人问鼎科甲。自然山水被人文情怀利用得如此淋漓尽致，实乃设计之巧思。

isting Cangpo Ancient Village was rebuilt by national architectural master Li Shiri in 1178 (the 5th year of Chunxi, Southern Song Dynasty) according to the layout of the "scholar's four jewels"(writing brush, ink stick, ink slab and paper)" in Chinese five-element and geomancy (Fengshui) theories. It has a history of 840 years.

The setting of Cangpo Ancient Village is according to the "scholar's four jewels" having the Bijia Mountain as a landmark. The straight east-west central street pointing directly at Bijia Mountain is called "Bijia Street". The nib of the pen is pointing at the west side of "Bijia Mountain" (also called "Wenbi Peak"), with the east and west ponds collectively called as "Yanchi Pond". The shadow of Bijia Mountain was reflected in the Yanchi Pond to form a Chinese painting-like landscape. The edges of the Yanchi Pond were constructed by placing dressed stones to the ink-stone slot. One big dressed stone was placed at the sides of the two ponds close to Pen Street, meaning ink-slab. The square-shaped village setting together with the farmlands in the surroundings, resembles a grid-shaped piece of Chinese art paper, expecting the future generations to have outstanding performance in the imperial examinations. The natural landscape has been used for expressing the cultural feelings, reflecting the concept of the design.

苍坡古村 Cangpo Ancient Village

岩头古村
Yantou Ancient Village

目的 Purpose	主要考察岩头古村的历史、建筑风格及艺术、与村落有关的人文历史 The field trip is mainly to study the history, architecture, art and associated culture of Yantou Ancient Village.
位置 Location	永嘉县东北约21.3千米 It is located at about 21.3km northeast of Yongjia County.
路线 Route	从永嘉县城经S223到达目的地，再沿村落内部道路游览 The village can be reached by S223 from Yongjia County. Tourism activities can be carried out along the village roads.

岩头古村是楠溪江中游最大的古村落，创建于五代末年，为金氏聚居地。到明世宗嘉靖年间（公元1522—1566年），由八世祖金永朴主持，进行全面规划修建，平面上略呈矩形。以科学的水利设施和巧妙的村庄布局而闻名，有着楠溪江中游规模最宏大、设计最精巧的村落供水系统和公共园林。设计者巧妙地利用村子西北高的地势，把水从西北引入村，再从村东南出。通过引水渠把山溪接入村子西北角，再分成几条支渠绕街傍户贯穿村舍民居，最后汇入村子东南部的几个人工池塘。

Yantou Ancient Village was established in the late Five Dynasties and became the settlement of the Jin family and the largest historical village in the Nanxijiang midstream. During the period from 1522 to 1566, it was comprehensively planned and reconstructed under the command of Jin Yongpu, eighth ancestor of the Jin family. The square village is known for its scientific water conservancy facilities and the perfect layout of the village. It becomes the largest and well-designed water supply system and public garden in the Nanxijiang midstream. By virtue of the northwest highland, it was designed by the designer to guide water to flow from the northwest to the south-eastern part of the village. The water flows around the houses through canals and enters the artificial pools located in the southeast of the village.

岩头古村之丽水街 Lishui Street of Yantou Ancient Village

芙蓉古村
Furong Ancient Village

目的 Purpose	主要考察芙蓉古村的历史、建筑风格及艺术、与村落有关的人文历史 The field trip is mainly to study the history, architecture, art and associated culture of Furong Ancient Village.
位置 Location	永嘉县东北约21千米 It is located at about 21km northeast of Yongjia County.
路线 Route	从永嘉县城经S223到达目的地，再沿村落内部道路游览 The village can be reached by S223 from Yongjia County. Tourism activities can be carried out along the village roads.

芙蓉古村始建于唐代末年，为陈姓聚居之地。正元年(1341年)重建芙蓉村，至今已有675年的历史。今芙蓉村基本沿袭原有聚落格局与规划面貌，平面上呈矩形。总体布局为“七星八斗”。星，指道路交会处的方形平台；斗，是水渠交汇处的方形水池，七星呈翼轸形分布，八斗为八卦形分布。道路与水系交叉纵横，成为完整的体系。和平时期用于生活的各个方面，防火、调节小气候、美化环境等。战争时期，“星”可以用来作指挥台，“斗”可以用来储水，以防火攻。

Furong Ancient Village was founded in late Tang Dynasty and became the settlement of the Chen family. It has a history of 675 years since its reconstruction in 1341. Following the original pattern of the village, the rectangular village features an overall layout of "seven platforms and eight pools". Platform refers to the square platform located at the intersection of the road, while pool refers to the square pool located at the intersection of the canal. Seven platforms are arranged according to the location of the Yi and Zhen constellations and eight pools are arranged based on the Eight Diagrams. The complete system composed of cross roads and rivers is used to prevent fire, adjust the microclimate and implement landscaping in peacetime. In times of war, the platform can be used to command the army, and the pool can be used for fire-fighting.

芙蓉古村之芙蓉亭 Furong Pavilion of Furong Ancient Village

考察须知

Field Trip Notes

野外考察注意事项

山林地区，可能会发生崩塌灾害，请随时保持高度警惕，在路面湿滑时、栈道陡峭拥挤处需格外小心。

科学考察请与地质公园管理部门联系配合工作。如在山林中迷路，请尽快拨打紧急救援电话。

不要破坏公园内的地质遗迹，保护公园自然环境。

Field Trip Precautions

Due to potential landslides in mountains, please be alert at all times. Please take extreme care to slippery tracks and crowded condition of the boardwalk.

Please make contact with the geopark management for scientific studies. If getting lost in mountains, please call the emergency rescue number.

Please help to protect the geological heritages and the natural environment of the geopark.

交通及相关信息

公园交通状况良好，确保了公园内各景观点的可达性。详细乘车信息请咨询雁荡山旅游公交公司，电话号码：0577—62157199

Traffic and Relevant Information

The good traffic conditions in the Geopark ensure the accessibility to scenic spots in the Geopark. Please contact Yandangshan Tourism Bus Company for more travel information. Tel: 0577—62157199

雁荡山世界地质公园联系方式

地址：浙江省乐清市雁荡镇雁山路88号

网址：http://engeopark.wzyds.com/

电话：086-577-62180067

传真：086-577-62180016

Contact Information of Yandangshan UNESCO Global Geopark

Address: No.88, Yanshan Road, Yandang Town, Yueqing City, Zhejiang Province

Website: http://engeopark.wzyds.com/

Tel.: 086-577-62180067

Fax: 086-577-62180016